START
WITH
SOIL

JULIET
SARGEANT

START
WITH
SOIL

SIMPLE STEPS FOR
A THRIVING GARDEN

CONTENTS

Introduction   6

1. WHAT IS SOIL AND WHY IS IT IMPORTANT?   10
How Soil Works: Soil Structure, Function and Biodiversity   12
Soil and Climate Change   24

2. GETTING TO KNOW YOUR SOIL   38
Soil in Our Gardens   40
How to Test Your Soil   48
Soil Through the Year   62
Soil in Our Homes   64
Soilless Gardening   66

3. GARDENING FOR BETTER SOIL   70
Compost and Mulch   72
The Case Against Improving Your Soil and Feeding Your Plants   84
Protecting Your Soil   92
No-Dig Gardening   100
Rain Gardens   110

4. WHAT TO PLANT FOR A HEALTHY SOIL   118
How Plants Help the Soil   120
Top Soil-Saving Plants   126
Plants for Different Soil Types   127

Soil Trouble-Shooting   142
Glossary   152
Index   155
Picture Credits   159

INTRODUCTION

Before I travelled to visit my extended family in East Africa, I had never seen a land so deprived of plants, and so began my journey to a better understanding of soil.

My father's family originates from Ukewere Island in Lake Victoria, and many of my aunts, uncles and cousins remain there, living as subsistence farmers. They grow, themselves, the food that they eat. If the crops don't grow, they don't eat.

Ukewere is the largest island in the largest lake in Africa, but few tourists venture there. Back in the 1960s my father considered it too remote and difficult to take his new bride from London, so my mother has never been. Even when I visited Tanzania in the 1990s I was discouraged. So, it took until the 2000s for my distant family to finally meet the 'doctor–turned–gardener' they were curious to know. I travelled with my husband and daughters. At Mwanza on the southern shore of the lake, we were warmly greeted by my uncles and escorted on the two-hour ferry journey across the fresh waters.

Nansio is the capital of Ukewere. The ferry docked on the vegetated shoreline of the town and we disembarked to walk down the wide dirt road to our hotel. Having just travelled from the foothills of Mount Kilimanjaro, here I was struck by the broad flatness of the island. The soil is that signature orange-red of so many African regions and, giving it little thought, I assumed it had the same richness of the Kilimanjaro volcanic farmlands.

Many people came to see us and we feasted on tasty local fish, fruits and vegetables. We were honoured with the killing of a goat and it was only gradually that I became aware of the lack that was masked by my family's generosity. Conversations were slow, because most people speak a dialect, so interpretation was first into the national language of Swahili and then again into English, for us.

We walked the land of the family farm, which has supported a growing population for the last hundred years. It was then that questions started to form in my mind. Most of the soil was bare and the few plants growing seemed sickly. I questioned gently, not wanting to cause offence, but there was no hesitation in sharing with me the struggles they were having. The crops were failing.

The situation had been worsening for years and nothing had come of government promises to send agricultural experts to help. My family had no idea of the cause, but I had my suspicions. The soil on Ukewere is sandy. Bending to scoop a handful, I let the dry dust drain through my fingers and I imagined how the nutrients, so essential for plant growth, also drain out as the rainwaters wash through the land. Later tests confirmed that their soil is 'dead': depleted of nutrients and completely lacking in organic matter.

It was on receiving this news that the importance of soil care dawned on me. *This silent catastrophe must be happening all over the world.* But something else also occurred to me. *I have my own soil at home. Am I doing the right things to look after it?*

I live in Sussex, in the south of England and I have the privilege of being able to pop to the supermarket if I run out of vegetables. My gardening is for my pleasure, not my survival. My motivation is for bigger flowers and tastier treats. However, I do think that there is something that I, you, we can contribute to the bigger picture as well.

No garden is an island; Nature does not recognise land registry and the ownership boundaries that we erect. Your unique piece of land is part of the 36 billion acres on earth – an integral part. What we do as home gardeners matters. We can make a difference, not just to our own well-being, but to the survival and health of our environment and our species.

It is difficult to say how much of the Earth's habitable land is garden, but to give you an idea, a quarter of the area of London is cultivated privately and similar spaces, taken as a whole, cover a greater area than all the UK nature reserves put together.

On its own, your small plot may be fairly insignificant, but as part of the whole it is important.

In this book, I hope to give you the tools to look after your soil. I am going to keep it practical and uncomplicated, so you can get going quickly on improving the way that you garden.

At the same time, I hope to pique your curiosity. The world of soil is still very much uncharted by scientists; we know so little about how it all works down there, beneath our feet. What we do know is that it is mind-boggling in its complexity and beautiful in its intricacy. Elements cycle efficiently between atmosphere, soil, water and plants. Nutrients are drawn to where they are needed. Previously unknown microorganisms are essential to larger life forms. Water and carbon are held in a fine balance, which is all too easily disrupted by our unthinking interventions.

Let's dig deep to find a better understanding of that much-neglected stuff upon which all life on land depends.

PART ONE:

WHAT IS SOIL AND WHY IS IT IMPORTANT?

HOW SOIL WORKS: SOIL STRUCTURE, FUNCTION AND BIODIVERSITY

Soil supports 95% of all global food production.

INTRODUCTION TO SOIL

Earth, mud, dirt, soil ... what is this stuff?

I think it is different things to different people, depending on how you want to use it. If you were making pottery or bricks you'd want a clay with a rich colour that's easy to work and heats well throughout. A civil engineer needs to understand the properties of the local soil that sits beneath a road or railway. A child plays in the squelching coolness of a mud pie and afterwards the parent scrubs the dirt from her clothes. We, as gardeners, want our soil to support ... life.

People agree that soil is the surface of the Earth and consists of water, air, minerals (solid inorganic substances) and organic matter (plants, animals and their faeces). However, some would say that the organic matter that is part of soil is only the dead, decomposing matter. The living minibeasts and microorganisms are essential to soil and part of a soil network, but are not actually soil. So, strictly speaking, it is not correct to say 'soil is alive' or we need to 'feed our soil', but I think it does help us to focus on the fact that all life on land is inextricably linked to the soil. The life forms within soil, which will become part of the soil – they are alive. Microorganisms and minibeasts feed, drink and breathe. For soil to be healthy, they must be able to live; and for them to live, the soil must be healthy.

The soil network is a term which describes the complex system of relationships between the different organisms which live in and on the soil. Soil is by far the most biodiverse (containing a variety of plant and animal life) material on the planet. A square metre of good soil can contain about a billion organisms, which can be divided by size into mega-, macro-, mesa- and micro-fauna, followed by tiny microscopic 'protists', a group of mostly single-celled organisms that includes some types of bacteria, slime moulds and algae.

A quarter of all land-based species inhabit the soil and it supports 95% of all global food production. But there's more – the Earth's soil cleans and filters water, as well as managing it to reduce flooding and recycle rainwater. It also cycles essential gases and nutrients, as well as locking up more carbon than our trees do, helping to manage global temperatures and climate change.

In our own gardens, a healthy soil will give us better crops and happier plants. By tending to your garden soil, you will help to make your plants more resilient to adverse weather, pests and diseases and climate change. You will notice an increase in biodiversity in your garden, which will then in turn improve your soil.

You may feel that your garden is but a small part, but it is a unique part. Every square metre of soil has its own individual character, structure and soil inhabitants.

The land that you tend is an essential piece of your local environment and your local environment is an essential part of the global whole. All interconnected.

SOIL FORMATION

Before plants colonised land, the Earth was covered in barren rocks, with some bacteria. There was more carbon dioxide (CO_2) and less oxygen in the atmosphere than there is now, but a particular type of bacteria, 'cyanobacteria', were able to survive by photosynthesising, drawing CO_2 from the air and releasing oxygen. These cyanobacteria then started to form mutually beneficial relationships (symbiosis) with other simple life forms, such as algae and lichen. In the meantime, physical and chemical weathering broke down the rocks into smaller and smaller pieces. This mush of rock, along with the decomposing organic matter of the early bacteria, algae and lichens, started to form soil. A mix of water, air, minerals and organic matter. As the soil became deeper, it was able to support the evolution of larger and more complex organisms. These, in their turn, died and decomposed, thus creating a system of regenerative soil formation which continues to this day.

Most of the organic matter in soil derives from plants, but animals, too, will die and decompose, releasing their nutrients back into the soil. Thus, the soil comes to contain a rich cocktail of compounds and elements, which can be used for plant growth.

This is a general description of early soil formation on Earth, but this process is continuing; soil is always forming and being lost.

ORGANIC VERSUS MINERAL SOIL

Most soils are a mixture of organic material and minerals, as well as water and air. The proportion of organic to mineral content will vary according to how and where the soil has formed, and it is this proportion that determines whether the soil is classed as 'mineral soil' or 'organic soil'.

If more than half of the top 80cm/32in of a soil is organic matter, or if there is a thick layer of organic matter resting directly on top of the underlying or 'parent' rock, the soil is classed as organic. A common example of this would be peat soil, which is derived from living organisms (plants in the case of peat) dying and decomposing to form thick organic and carbon-rich layers.

A mineral soil comes from weathered rock, with much less laying down of organic plant and animal matter. The minerals either form in place from weathering (residual soil) or are transported, often over a great distance, by wind or water, and then laid down (sedimentary soil).

In the UK, there are over 700 different types of soil and most of them are mineral. So unless you know that you live in a peaty area, your soil is probably mineral.

Below: Soil particles of clay, silt and sand group together to form aggregates of varying sizes and between the interlocking aggregates are pores.

PARTICLES AND AGGREGATES

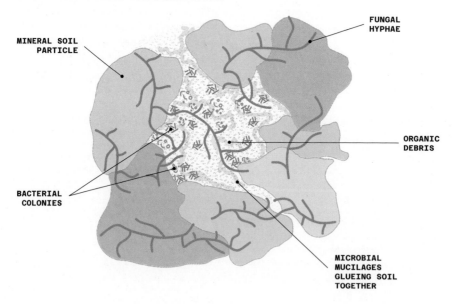

FUNGAL HYPHAE

MINERAL SOIL PARTICLE

ORGANIC DEBRIS

BACTERIAL COLONIES

MICROBIAL MUCILAGES GLUEING SOIL TOGETHER

STRUCTURE

The structure of a soil is the way that the particles are arranged and held together.

Clumps of organic and mineral particles are known as aggregates or peds, and the air pockets between the particles are known as pores.

There are several physical, chemical and biological factors that bind the particles together. Tiny electrical charges draw particles towards one another and also draw a thin film of water around their surface. Plant roots and fungal hyphae (thread-like parts of fungi) that grow through the soil, as well as 'biological glues' exuded from bacteria, fungi and plant roots, help to hold everything together.

The pores create a network of spaces that allow water and gases through. The larger the pores and the more connected they are, the more porous or permeable (penetrable) the soil.

Aggregates are grouped by Type (affecting permeability), Class (aggregate size) and Grade (aggregate stability). There are several aggregate types, but the most useful for us gardeners are the granular types, which are most common, where organic matter is present and generally good permeability. A 'platy' aggregate type is found where soil is compacted and has low permeability.

A good soil structure will have stable aggregates, which means they are bound together securely enough that they are not easily washed or blown away, with a good network of interconnected pores allowing air and water to flow freely through the soil (good permeability). Typically, this soil might be approximately 50% solid, 25% air and 25% water.

The air and water held reliably within the soil allow for life to thrive, increasing biodiversity. The increase in life then adds to a positive cycle where roots and minibeasts move through the soil, aerating it and stabilising aggregates with their exudates (sticky plant or animal excretions).

Thus, a good soil structure supports biodiversity and agricultural and garden productivity. But there are more benefits: a healthy soil has enough pores to act like a sponge when it rains. The rainwater can travel vertically down through the soil and can also be held temporarily between aggregates. When the rain stops, the soil gradually releases the water back out into the wider environment. Soil also acts to filter water and reduce floods; it is a massive natural water management system.

Our soil is also the largest store of carbon on Earth (larger than the rainforests); it holds twice as much carbon as the atmosphere. By locking away vast stores of carbon, preventing them from entering the atmosphere, soil helps to slow climate change.

Soil structure is fragile and can easily be destroyed by farming practices such as tilling; working in cold, wet weather; soil compaction by machinery and livestock; flooding; and erosion from the removal of trees and hedges. On a smaller scale in our own gardens, we should avoid deep digging, working the soil in winter, compaction and overwatering.

We can improve soil structure in our gardens by planting trees and hedges (and in fact any plants), adopting 'no-dig' techniques with application of organic matter, mulches and cover crops to the soil surface, and encouraging biodiversity.

TEXTURE

The weathering of rock results in particles of different sizes called sand, silt and clay. The proportion of large particles to small gives a soil many of its characteristics and its descriptive name. This varying proportion of particle sizes is referred to as the 'texture' of the soil.

Sand particles are the largest and are felt in the hand as gritty. If a soil has a high proportion of larger sand particles, its texture will be light, with air spaces between the particles. Water will move freely between the sand particles, allowing it to drain freely. As the water runs through the soil, it will tend to take soluble nutrients with it, so often a sandy soil is light, dry and poor in nutrients. If you take a handful of sand and let it run through your fingers, you will understand why, according to the Bible, it is a 'foolish man' who builds his house on sand.

Clay particles are the smallest and tend to be plate-like in shape. These microscopic plates stack and slide over each other, resulting in fewer air spaces between them. So, the soil is heavy and it is difficult for water to move between the particles and through the soil. Therefore, nutrients are often 'locked up' in the soil: there for the taking, but difficult for the plant roots to access. Clay is also a very stable soil, beloved of the construction industry.

Silt particles sit between sand and clay in size. Silt can easily be mistaken for clay, but it has a shinier appearance when wet and none of the structural stability of clay.

The texture of a soil can either be the same or change as you go deeper underground. The change can be gradual or sudden from one texture to another.

Remember texture refers to the type of soil particles, whereas structure refers to

SAND, SILT AND CLAY

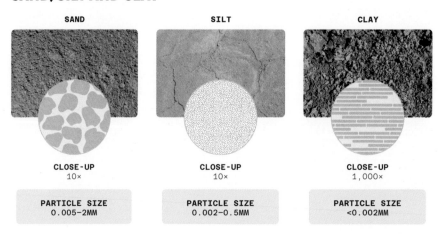

SAND	SILT	CLAY
CLOSE-UP 10×	CLOSE-UP 10×	CLOSE-UP 1,000×
PARTICLE SIZE 0.005–2MM	PARTICLE SIZE 0.002–0.5MM	PARTICLE SIZE <0.002MM

how those particles are grouped together (in aggregates) and how strongly they stick together. Texture is pretty much fixed, but with good garden practices you can improve your soil structure.

SOIL PROFILE

Soil takes a long time to form: on average, one hundred years for one inch. As soil forms, it is laid down in layers, which can often be seen in the 'soil profile' – a vertical section of a particular soil. To view a soil profile you dig straight down, trying not to disturb the sides of the hole you are digging.

Each formative layer of the soil profile is called a horizon. There are several named horizons, but for the gardener, the most important are the 'O', 'A', 'B' and 'R' horizons.

The 'O' (organic layer) horizon is the uppermost surface of the soil. It is here that fresh leaves and decaying organic matter can clearly be seen. There may be stones, rocks and twigs. This part of the soil is rich in organic matter and so is often quite dark in colour.

The 'A' (mineral layer or topsoil) horizon is the topsoil where most of the plant action takes place. It is typically only 2–8cm (¾–3¼in) deep, but it can be as much as a metre. Topsoil is rich in organic matter, although this is still only about 5% of it; 45% is minerals, 25% water and 25% air.

Here, seeds germinate and the roots of the plants press through the soil. As the roots grow, they push soil particles aside, creating air pockets, which remain when the plant dies and the roots rot away.

'A' is a very busy place, teeming with life, both microscopic and macroscopic. It is often said that there are more microorganisms in a handful of soil than there are people on earth.

Earthworms inhabit the uppermost layers of the soil. They are fundamental to the health of the soil and also a good indicator of it. As a worm travels through the soil, bodily fluids exuded from its surface help to stick soil particles together as more stable aggregates. Its burrows aerate the soil and create tracks for water to move through. Worms also help incorporate organic matter into the soil, as they drag leaves down for food and then digest nutrients and excrete their waste.

The 'B' (subsoil) horizon is the layer formed from the underlying bedrock in areas of mineral soil, and it contains less organic matter, so is a lighter colour than topsoil. Water is often held in the higher clay content of subsoil, and minerals leached from the topsoil above may collect there. In an ideal world, the 'B' and 'A' layers would not mix, but ploughing and deep cultivation can result in disruption of the soil profile.

The 'R' horizon is the bedrock.

In your garden soil, you may not be able to see clearly divided horizons. They may have been mixed up over time, or possibly were never there in the first place. Horizons, their number and order within a soil profile can vary greatly. However, it is worth digging a profile pit, in order to understand more about the make-up and history of the soil on your plot.

THE PLANT RHIZOSPHERE

Plants (and cyanobacteria) are the only organisms that can photosynthesise, drawing carbon from the atmosphere and getting their energy directly from the sun. The rest of us are dependent on plants for our energy and for carbon, which is the building block of life. When we (or other animals) eat plants, we are ingesting the energy they have captured from the sun in the form of sugars. And we are also ingesting carbon molecules in the form of sugars and proteins in plant leaves, fruit, flowers and seeds.

Plants absorb other vital nutrients from the soil through their roots. Again, we animals have no way of getting these nutrients except by eating plants, or eating other animals which have eaten plants. Vitamins, minerals, carbon and energy all come from plants one way or another and the plants get them ... from the soil and the sun.

When I was at school, I was taught that a plant sucks water up through its roots and dissolved in that water are the essential nutrients. The reality is much more complex and amazing than that and to this date still little understood.

The rhizosphere is contained in the few millimetres of soil that directly surround the roots of a plant. I think of it as the fingers of a glove encasing the roots, their secretions and the microorganisms that are influenced by, and influence, the growth of the plant. This bespoke, beneficial world that the plant creates for itself is called the 'root microbiome'.

Over millions of years, plants have evolved symbiotic relationships with bacteria and fungi, which inhabit plant roots to varying degrees of proximity, some reaching inside the root between cells and others actually invading the root cells themselves.

Plants cannot move quickly and easily to find food, as animals can. They have to rely on either growing slowly towards nutrients (root foraging) or attracting other beneficial organisms that will help them. Plant roots release 'rhizodeposits', shedding cells, mucilage or chemicals that attract bacteria and fungi. In the rhizosphere there is an extremely high concentration of nutrients and thus microbes – far higher than in the general soil. It is a crowded and competitive space.

The three main rhizosphere-dwelling organisms that we know the most about are rhizobia bacteria of legumes (the pea plant family), mycorrhizal fungi and plant growth promoting rhizobacteria (PGPR).

Nitrogen is an essential nutrient for plants. Rhizobia live in lumps, or 'nodules', on legume roots as a response to low nitrogen levels in the soil. The bacteria convert gaseous nitrogen to soluble ammonia, which the plants can then absorb and utilise.

Fungi cannot photosynthesise, so mycorrhizae (fungal roots) live in and around the roots of plants to access the carbon sugars produced by the plant. In return, the fungi increase a plant's root surface area for mineral absorption, and donate water, phosphorus and other essential micronutrients.

PGPR are attracted to the plant root rhizodeposits. They colonise the root surface and fight pathogens, release growth hormones and aid nutrient uptake.

Beyond these beneficial relationships, the plant can help itself by altering the chemical composition of the rhizosphere to improve nutrient uptake. For example, nitrogen is abundant in the air, but not very available to plants in the soil. A plant can alter the pH around the root zone in order to increase available forms of nitrogen.

Similarly, phosphorus is generally insoluble in soil and tightly bound to other molecules. Plants send out organic acids to dissolve phosphate and so absorb it.

These are just two of the many nutrients which a plant needs to thrive.

THE RHIZOSPHERE

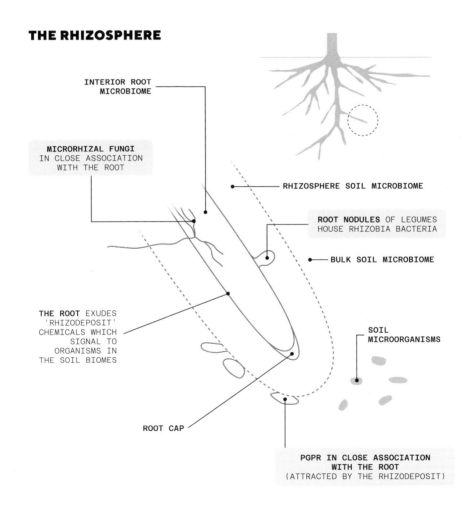

INTERIOR ROOT MICROBIOME

MICRORHIZAL FUNGI IN CLOSE ASSOCIATION WITH THE ROOT

RHIZOSPHERE SOIL MICROBIOME

ROOT NODULES OF LEGUMES HOUSE RHIZOBIA BACTERIA

BULK SOIL MICROBIOME

THE ROOT EXUDES 'RHIZODEPOSIT' CHEMICALS WHICH SIGNAL TO ORGANISMS IN THE SOIL BIOMES

SOIL MICROORGANISMS

ROOT CAP

PGPR IN CLOSE ASSOCIATION WITH THE ROOT (ATTRACTED BY THE RHIZODEPOSIT)

SOIL NUTRIENTS

The nutritional requirements for a plant to thrive are complex and many, but there are a few basic principles which will help you to make the best decisions for your garden, its plants and soil.

Plant nutrients are traditionally divided into primary macronutrients (nitrogen, phosphorus, and potassium), and secondary macronutrients (calcium, magnesium and sulphur), which plants need more of, and micronutrients, which they need in minute quantities.

Most nutrients are readily available in a healthy soil, but there are a few common deficiencies which you may come across and which can usually be remedied by adding organic matter. You may need a specific fertiliser, in which case opt for an organic preparation if you can.

If your soil is healthy, you are unlikely to have nutritional problems that require testing, but if you do, the Royal Horticultural Society has a soil analysis service. They will give you a profile of the soil nutrients and organic content of your soil and also advise on any measures that you could take if you are looking to grow particular plants or crops.

Nitrogen is abundant in the atmosphere, but in soil it is often either washed away with rain or bound up in inaccessible forms. Nitrogen is essential for developing the chlorophyll used in plant photosynthesis. Deficiency after a wet winter is a common cause of yellowing leaves and stunted growth in spring. To resolve this, apply generous amounts of organic matter and possibly a slow-release fertiliser, such as fish, blood and bone.

Phosphorus is essential for healthy root growth. Deficiency in soil is unusual and can be remedied with fish, blood and bone.

Potassium, or potash, is used by the plant in flower, fruit and seed growth. It can be deficient if your soil is very light, sandy or chalky, without the organic matter to hold on to moisture and nutrients (clay soil is usually rich in potassium). Organic sources of potassium include seaweed preparations, banana peels, manures, comfrey tea and wood ash.

If you have been fertilising regularly with a high potassium feed, you might inadvertently cause another problem: magnesium deficiency. This can cause leaf yellowing of older leaves in tomato plants and crops that you have been feeding to improve fruit or flower yields. The plant's enthusiastic uptake of potassium ions (electrically charged atoms that aide plants to absorb nutrients) can displace their uptake of magnesium. This can also be a problem on a light sandy soil, where magnesium levels might be low or washed out by winter rains. A quick-fix is Epsom salts, but for the long term apply organic matter and general garden compost regularly.

A common problem for acid-loving (ericaceous) plants is yellowing of the leaves due to iron deficiency. This is not because there is low iron content in the soil (this is rare), but because a high soil pH is preventing the plant from absorbing the iron that it needs. It is better not to try and grow acid-loving plants on alkaline soil, but if that horse has already bolted, then mulching with acidic organic matter and feeding with a chelated iron seaweed fertiliser will hopefully help.

Normally, you will not need fertiliser if you keep your soil healthy, but the fertilisers that you might find most useful would be:

- Fish, blood and bone
- Seaweed preparations
- Homemade comfrey, nettle or compost teas

Below: Soil nutrient deficiencies often show in the leaves of plants. Yellowing of the leaves (chlorosis), such as with iron deficiency, is a common symptom.

SOIL BIODIVERSITY

Earth's biodiversity is essential to our food security and survival. 40% of organisms in the terrestrial ecosystem have a life cycle which involves soil and 25% of animal species actually live underground. So, the soil is a major global reservoir of biodiversity, holding more microorganisms in a teaspoon than there are people on the whole Earth.

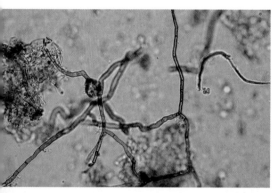

Most soil organisms cannot be seen with the naked eye. They include nematodes, bacteria and tardigrades, which decompose organic matter and recycle its nutrients.

Then there are protozoa which swim in the water in soil pores and the watery film over soil particles. Protozoa consume bacteria, releasing nitrogen in a form which can then be taken up by plants. Some bacteria also help to deliver nutrients to plants by fixing nitrogen to make it available to the plant roots.

Fungi have a fine network of root-like threads called mycelia, which spread through the soil. Mycelia form close symbiotic relationships with plant roots, and in doing so, they link the plants together in a widespread mutually beneficial community. We now know that through this Common Mycelial Network plants communicate with each other, sharing nutrients and also information about the local environment. Fungi have been shown to increase pest and stress resilience in plants. The fungal mycelia transport nutrients (nitrogen and phosphates) around the soil and into plant roots and, in return, the plants supply the fungi with carbon in the form of sugars, which the fungi cannot make themselves because, unlike green plants, they cannot photosynthesise.

The beneficial effects of fungi do not end there. The physical presence of their hyphae knits the soil together, reducing erosion. They are essential in decomposing organic matter, in particular woody material, and so are part of the process of adding organic compounds to the soil. Fungi can filter heavy metals and pollutants.

Unfortunately, these beneficial fungi are also susceptible to fungicides and artificial fertiliser, so when we apply chemicals to our soil, we may be killing or reducing the actions of these essential soil organisms.

The most well-known inhabitants of soil are the many varieties of earthworms. Each has a favourite layer of the soil to inhabit and there they burrow and eat decaying plant matter. They drag leaves and other plant debris down into their tunnels, burying it and turning it into fine worm casts (soil that has passed through a worm), which fertilise the soil.

Each minibeast, macro or microorganism present in soil, from the woodlouse scurrying through the leaf litter to the tiny springtail beneath, has a job to do. They are all part of the complex and delicate web of life in the soil that constantly recycles the Earth's resources for their benefit and ours.

Below: A woodlouse in soil.

SOIL AND CLIMATE CHANGE

Soil is a reservoir of carbon, nitrogen, phosphorus and water: releasing, absorbing and storing them to maintain a balance with the atmosphere.

INTRODUCTION

Our world, with its atmosphere, is basically a closed system with compounds and elements constantly being recycled. The many processes are held in fine balance, different cycles interweaving and supporting each other like the many links of a chainmail suit of armour.

Many of these biogeochemical cycles loop between sky, soil and sea in an unending flow, refreshing essential supplies required for the survival of life on Earth.

Soil acts as a giant reservoir for carbon, nitrogen, phosphorus and water, releasing and also absorbing and sequestering it (hiding it away). The soil organisms described in the last section help regulate this process.

We now know that many of our human activities have detrimental effects on these cycles. This also goes for other physical, chemical and biological systems in soil.

SOIL EROSION BY WIND AND DEFORESTATION

You have probably seen wind whipping up dust along a dry road in summer and perhaps you have heard of the American Dust Bowl: in the 1930s vast tracts of native prairie land were ploughed up to grow wheat during the First World War, and then a drought came. The ploughing churned up the soil's surface, leaving it dry and unprotected by plants. The topsoil then literally blew away, taking with it the soil's healthy structure and fertility.

In the UK today, erosion (wind and water) moves more than 2.2 million tonnes of soil every year – often into our water courses. This loss of valuable fertile topsoil is responsible for the loss of millions of pounds of productivity.

The effect of wind depends on several factors, including the speed and direction of the wind itself, but also the size of the soil particles, their dryness and their aggregation into small, heavier clumps. The soil particles need to be of a certain size and weight to be carried off.

A healthy soil has enough organic matter to bind the particles so that they stay grounded. So, organic matter is protective, as is a good covering (at least 30%) of plant ground cover and also trees and hedges which act as windbreaks.

To protect our topsoil from wind erosion, both in agriculture and at home, we can plant and maintain trees and hedges. Their leaves and branches act as windbreaks, and their roots help to bind the soil in place. The trees and shrubs also drop leaf litter, which decomposes into organic matter, increasing soil particle aggregation. This aggregation

Opposite: Tangential wind lifts light, dry soil particles into the air.

Above: Unimpeded rainwater creates gullies through unvegetated terrain.

is also assisted by bacteria and fungi (see page 15).

The presence of larger plants and their decomposing leaf litter creates fertile soil for the establishment of groundcover plants beneath them, and these in turn help stabilise the soil further.

So, put simply, if you want to protect soil from blowing away, plant more plants!

FLOODING

The story for flooding is a little different. Between soil particles there are air pockets or pores of different sizes. There are also larger spaces formed by physical movement of the soil or by the activities of animals burrowing and earthworms tunnelling. These pores create a sponge-like structure which gives many soils a permeability that enables them to soak up rainwater. Thus, the soil can hold on to, or lessen the damage of, a downpour for a short while and then gradually, under gravity, slowly release the water out into the environment or down into aquifers (underground watercourses) in the lower soil layers.

Less permeable soils, like heavy clay or compacted soils, have fewer pores and so cannot act as sponges. A downpour of rain will tend to run off the surface of the soil and possibly into rivers, potentially taking soil particles, contaminants and fertilisers with it. Sometimes, the water runs so quickly off the surface that the soil layers beneath can remain quite dry.

Organic matter helps to stick soil particles together into small, stable clumps or aggregates. It is between these soil aggregates that you will find a healthy variety of pore sizes to create the desirable sponge-like structure. So, again, organic matter helps keep a soil healthy and mitigate the damaging effects of rainwater. Such a soil will have about 10–25% air capacity.

Subsoils naturally have less organic matter and so tend to have less water permeability. However, water can track down through subsoil via burrows and fissures, often eventually reaching the aquifers beneath.

If a soil becomes flooded and saturated, the air-filled pores are filled with water. If this continues for too long, then beneficial aerobic bacteria do not have the air that they need for their biological functions. Anaerobic (oxygen-free)

Above: Standing flood water reduces oxygenation of the soil and can cause soil organisms, and later plants, to die.

bacteria predominate and that is when you might smell sulphur or a sourness to the soil. This increases toxic compounds in the soil, upsets the pH balance and decreases the availability of beneficial nitrogen to plants. Plants cannot survive in this situation, their leaves will yellow, and wilt, the roots will rot and eventually the plant will die.

Beneficial fungi are also lost, decreasing the decomposition of organic matter, which in turn reduces the aggregation of the soil into healthy clumps. So, soil chemistry, biology and structure are all disrupted by the presence of long-term water within the soil pores – or flooding.

Above: Repeated compression of the soil can cause compaction and damage to the structure and aeration.

SOIL COMPACTION

The importance of the pores within soil is also apparent when we look at the effects of soil compaction.

This is when physical weight or repeated tracking across an area of soil literally squeezes the air out, pressing the particles together, creating an unhealthy anaerobic environment. In such an environment anaerobic bacterial activity takes over from beneficial aerobic reactions, creating a toxic environment detrimental to plant and animal survival. The soil surface may crust over and you may notice standing water and the proliferation of tell-tale weeds, such as creeping buttercup or Yorkshire fog grass.

A soil's texture has an effect on its susceptibility to compaction. Sandy soil has a large proportion of larger soil particles. It generally has a weak structure and can easily be eroded by water or wind but not so much by weight.

Silty soil is easily compacted and can be damaged by working it when it is wet. It can be prone to capping, which is when physical action causes the soil clumps to disaggregate. They break apart, releasing the smaller soil particles, which then block air pockets, creating a seal or 'cap' across the surface of the soil.

Clay particles are the smallest and can very easily be compacted, damaging the soil structure. Clay drains less easily than other soils and can lie wet for long periods. Even if the surface appears dry, the subsoil underneath may still be wet, particularly in spring, so beware of working the soil too soon after a cold, wet winter. Clay is easily compacted and in summer it can bake hard, often too hard to work.

It is difficult to alter the soil type or texture that you have in your garden; however, with good gardening habits, it is possible to mitigate the characteristics of that soil and improve its structure. You can manage your soil to help it withstand the physical onslaughts of water, wind and weight.

You will not, by now, be surprised to hear that all the soil types above can be helped with the application of organic matter. It improves soil structure, creating stable aggregates of soil particles which are less prone to physical compaction. Deep-rooted plants, such as clover, chicory or lucerne, can help to break up the soil, and green manures (crops planted and then dug into the soil) will protect its surface from water run-off.

It is also important to protect your soil, not allowing heavy vehicles to track across it repeatedly if, for example, you are having building work done, especially when it is wet and cold, and to avoid leaving heavy items for any length of time in the same spot on a growing area.

PEAT

Peat soil has a high content of organic matter which has decomposed over many years (often thousands). The conditions are usually waterlogged and acidic with low oxygen levels (like in a bog), so the plants do not fully decompose, giving the soil an open structure which is particularly good at absorbing and holding on to moisture.

When you hold peat in your hand, it feels light, moist and spongy, with a fresh, earthy smell to it. The high organic, plant content of peat means that it is high in carbon. For that reason, it has been used for hundreds of years for burning to produce heat and power. It has also been used by generations of gardeners to

improve soil by adding it as organic matter to open up the structure and increase the water-holding capacity and nutrients of their plots.

The acidity of peat also means that if you want to grow acid-loving plants, you could either add peat to your native soil or grow directly into peaty compost from the garden centre.

However, peat forms so slowly, under very specific circumstances, that it is essentially a non-renewable resource. It

Above: Large-scale harvesting of peat.

Right: Peat bogs are rare, ancient and biodiverse habitats.

takes ten years for just 1cm/½in of peat to form. Therefore, the large-scale extraction of peat is increasingly criticised.

As well as the direct loss of the peat itself, peat extraction causes structural soil damage and disruption of drainage and decreased water filtration in some of the rarest landscapes on Earth. It also causes loss of ancient habitats and biodiversity in delicate ecosystems. The high carbon content of peat remains locked into the soil, acting as a vast carbon sink. If we extract the peat, we release that carbon into the atmosphere and spark reactions in the soil which reverse the sequestration of carbon in future years.

It is for these reasons that the UK government has set out a timeline for the ban of all uses of peat by 2030, the professional use in horticulture by 2026 (with exemptions) and amateur use from 2024.

In the garden centres now, you will find a number of peat-free composts available.

NITROGEN POLLUTION

Nitrogen has a cycle between soil, water and air. Most of the Earth's nitrogen is in the air, which is 78% nitrogen. However, most non-plant living organisms cannot make use of this gaseous form, so we rely on plants. Before we can get the nitrogen we need from plants, it needs to be 'fixed' for them by nitrogen-fixing bacteria. Several types of bacteria are able to capture gaseous nitrogen to form ammonium and nitrates, which can be taken up by plants to form essential proteins. Once incorporated into the plant structure, the nitrogen becomes organic in form and so can be eaten by other animals or people to make important proteins such as DNA. This is known as biological nitrogen fixation and it is an indispensable step in the supply of nitrogen to most ecosystems, keeping them in a fine balance.

When an animal excretes, dies and decays, the nitrogen in its organic matter is converted back by microorganisms into inorganic soil nitrogen in several steps from ammonium to nitrites, or nitrates by bacteria, making it available to be taken up by plants.

Alternatively, the nitrates in the soil can be converted by the action of microorganisms to nitrous oxide or nitrogen to be released back into the atmosphere (denitrification). And ammonium can also be converted to gaseous ammonia (volatilisation). Denitrification can increase as a result of soil flooding or compaction, and this increases bacterial anaerobic activity.

Both ammonium and nitrate are forms which can be taken up by the root hairs of plants. However, wheres ammonium is relatively stable and fixed in the soil, nitrate is soluble and moves easily through the soil (leaching).

With the Haber-Bosch process invented in the 1940s, nitrogen can be industrially combined with hydrogen to create chemical fertilisers. This introduces nitrogen into the nitrogen cycle, artificially increasing the available forms of nitrogen. This has resulted in increased crop yields, but it has also caused a global problem with excess soluble nitrates leaching out of the soil and into rivers and seas. The high nitrogen levels in the water cause algal blooms, which shut out light and reduce oxygen in the water, with detrimental knock-on effects.

High nitrogen levels in soil are also thought to be contributing to increasing levels of nitrous oxide (N_2O). In the atmosphere, N_2O can break down to form other molecules that destroy the ozone layer. It is the ozone layer that protects the planet from ultraviolet radiation. Nitrous oxide is less talked about than carbon dioxide as a greenhouse gas, but it persists in the atmosphere for decades and is more efficient at trapping heat around the Earth. There is also evidence that artificially inflated nitrogen levels in soil may switch off the nitrogen-fixing and nitrifying microorganisms in the soil, thus interrupting the natural nitrogen cycle.

Right: Nitrogen occurs in different forms, which interchange and cycle through soil, air and living organisms.

NITROGEN CYCLE

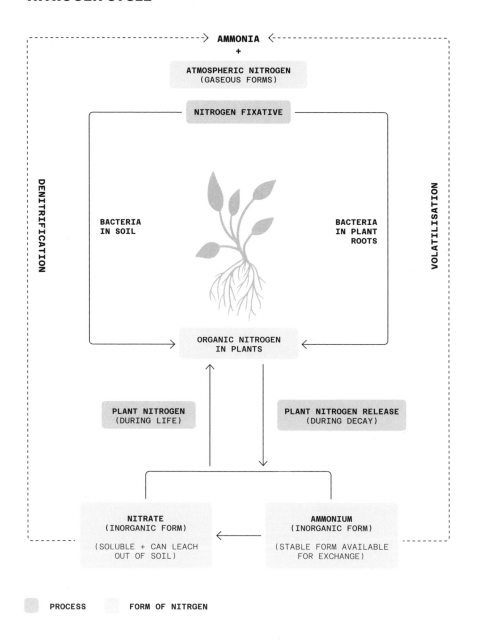

AMMONIA
+

ATMOSPHERIC NITROGEN
(GASEOUS FORMS)

NITROGEN FIXATIVE

DENITRIFICATION

VOLATILISATION

BACTERIA
IN SOIL

BACTERIA
IN PLANT
ROOTS

ORGANIC NITROGEN
IN PLANTS

PLANT NITROGEN
(DURING LIFE)

PLANT NITROGEN RELEASE
(DURING DECAY)

NITRATE
(INORGANIC FORM)

(SOLUBLE + CAN LEACH
OUT OF SOIL)

AMMONIUM
(INORGANIC FORM)

(STABLE FORM AVAILABLE
FOR EXCHANGE)

PROCESS   FORM OF NITRGEN

SOIL CARBON AND AIR CARBON POLLUTION

Carbon is a building block of life on Earth. It is a very versatile and abundant element found in all living things. It makes the cell walls of plants and animals. Our bodies are mainly carbon with oxygen. As well as this living organic form, non-living organic carbon can be found in the soil as decaying plants, animals and microorganisms or 'humus'.

Inorganic carbon atoms can be found in the atmosphere, combined with oxygen atoms, as carbon dioxide. Or, they can also be locked into soil in relatively inactive states of solid carbon - such as fossil fuels and diamonds.

The carbon atoms themselves are the same; they are just cycling round and round from organic to inorganic and in different states of each.

If a cycle can have a beginning, it all starts with plants. They take in carbon atoms as carbon dioxide during photosynthesis, and using energy from the sun, they make their own plant cells and also sugars, which store the sun's energy in useable packages. So, plants (algae and cyanobacteria) are the only organisms on Earth that can capture carbon from the air and also capture energy from the sun. Very clever scientists spend their days trying to find efficient ways to save our planet by extracting carbon from our atmosphere, whilst the plants outside their laboratory windows are silently doing just that and have been for millions of years!

So, imagine a plant, let's call it a blade of grass, which has carbon in its structure and also in the sweet sugar that you can taste if you chew on that blade. If a calf now eats that grass, the carbon transfers to the calf, either to make its cells as it grows or to give it energy. If humans then eat the cow, again the carbon is transferred.

If, however, the grass is not eaten but dies (or, if for that matter the calf dies before it is eaten by us), the carbon is returned to the soil when it decomposes and its cells are broken down into organic matter in the soil. Organic carbon can then in time convert to inorganic by becoming fossil fuels or diamonds, or it can stay in the soil in a complex organic form called humus.

So far it has all been one way: carbon travelling from air down into the soil. But there is a reverse flow. Animals cannot photosynthesise; instead we breathe in oxygen and breathe out carbon dioxide, so if it were not for plants doing the opposite, our atmosphere would soon be depleted of oxygen and filled with CO_2. We would not be able to breathe and live.

Remember all that carbon trapped in the soil? In the UK that's an estimated 98 billion tonnes of carbon. But it can be released from the soil; if fossil fuels are burned, carbon dioxide is given off, and as soil microorganisms breathe, they too give off carbon dioxide. Ten per cent of the world's carbon emissions are stored in soil.

Our problem is that we have upset the balance of the carbon cycle by burning too many fossil fuels whilst reducing the number of plants (most importantly trees), and also disrupting the soil's ability to do its job of carbon sequestration.

Less well known than the effects of fossil fuels is the fact that deep digging and ploughing of soil aerates it, which encourages the microorganisms to go into overdrive, frantically respiring, breaking down organic matter and releasing more carbon dioxide. Our farming and gardening methods are contributing to carbon release, the greenhouse effect and climate change. However, in the agricultural sector the picture is extremely complex, so it is not currently clear whether farmers should stop ploughing or not.

What is agreed is the potential of our soil to capture and lock in much more of the damaging carbon dioxide than it currently is. This can be achieved by adding organic matter, protecting the structure of the soil and planting more plants.

CARBON CYCLE

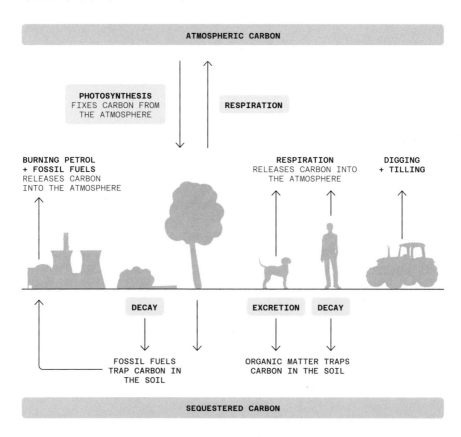

MICROPLASTICS

The problem of microplastics (MPs) in our soils and fresh water is now thought to be a greater problem than the much-publicised issue of marine plastic pollution.

One-third of the plastic that is dumped in landfill eventually works its way into our soils and waterways as microplastic particles and fibres. However, the main source of microplastic fibres is in fact from washing our clothes made from man-made textiles, which are shed into washing machines. Cosmetic microbeads similarly find their way into the environment.

Microplastics are present in human sewage and 80–90% of that plastic remains after treatment of the sewage in the sludge that is used by farmers to fertilise the soil of their fields. Microplastics have now been found in tap-water and, consistently, in human blood.

The microplastics can also carry chemicals on their surfaces, which can then leach out into the ecosystem. A worrying global cycle of potentially damaging artificial particles is now in place.

Unfortunately, for those of us who assiduously recycle our plastic in the hope of 'saving the planet', the recycling process also produces microplastics, and it has been shown the recycled polyester fabric in fact sheds more MPs than virgin polyester. However, the pros and cons of plastics, carbon footprint and recycling are complex and beyond the scope of this book, so I certainly would not want to put anyone off recycling.

So, apart from the unpleasant idea of having plastic balls circulating around our bloodstreams, why are MPs a problem? There is still a great deal of research to be undertaken to further our understanding of their effects, but so far it seems that MPs have mainly negative effects on natural processes.

In the soil we now understand that MPs can decrease soil aeration, increase soil erosion, alter pH, decrease the nutrients available to plants, decrease soil biodiversity and in most cases inhibit plant growth. They also seem to alter the burrowing behaviour of certain earthworm species and have been found to cross the blood–brain barrier of fish to affect their behaviour.

Right: It is becoming increasingly apparent that macro-and micro-plastics are not inert, as once thought.

CHANGING LANDSCAPES AND GARDEN AESTHETICS

We think of landscape as having a comforting permanence. It is the scenic backdrop against which our lives are played out. We are used to the colours and textures of our own particular corner of the planet and we like its familiarity. Beloved features of the landscape are woven into our cultures: art, music, myths and ceremonies. In England, the 'mighty oak' is a symbol of steadfastness, of unerring faithfulness and safe predictability. Silhouettes of ancient oaks have formed the backbone of the English countryside since living memory; our security rests on the certainty of their permanence.

Enter 'Acute Oak Decline' and 'Oak Processionary Moth': just two of the hazards now befalling our oak tree population and causing falling numbers of healthy trees. Soon, along with an increase in climate-change related pests and diseases of sweet chestnuts, ash trees, pines and the whole range of deciduous trees, our familiar treescapes will look very different.

For years now, scientists have been researching climate and disease resistant tree varieties, and already foresters and landscapers are changing their planting palettes. It is likely that the familiar trees of the European land-mass will become the new backdrop to the next act of 'The English Play'. The change is already happening, and obvious as it may sound, I do think it bears saying: change means that things will be different.

In our own gardens, even with every mitigating effort that we can make, we will see an increasingly unpredictable climate

Above: Familiar tree shapes and silhouettes are fast disappearing from landscapes.

Right: Climate change is responsible for many pests and diseases migrating and causing problems in established tree populations.

and its resulting effects on flora and fauna. With the recent devastating diseases of box plants, traditional topiary is becoming a thing of the past. Arguably, the 'English Garden Style' will soon be climatically and ecologically inappropriate. What will replace it? What is the English garden of the future? That is our project.

THE ROLE OF SOIL IN CLIMATE RESILIENCE

Soil is like a skin covering 30% of the Earth's surface. It is a vital part of maintaining the balance of the biogeochemical cycles of water, carbon and nitrogen that sustain life.

The soil manages the flow of water after rain, acting like a sponge to soak it up and then release it gradually back into the system. One acre of healthy soil can hold about 9,000 tonnes of water. The structure and water-holding capacity of soil is improved when it has a good amount of organic matter.

If there is a consistent flow of water delivered to the roots of plants, they can grow better and be healthy and more resilient to environmental assaults. The plants will grow well and their foliage cover and root infiltration in turn will protect the soil from erosion by water and wind. (In 2015, DEFRA estimated that about 3 million tonnes of topsoil are eroded in the UK each year.)

The roots of living plants and the decay of dead plant material increase the organic content of soil, so you can see a beneficial cycle is in play. As plants grow, they photosynthesise, capturing carbon from the carbon dioxide in the air and locking it away first as sugars in their leaves and then as sugars and structural components in the soil. Soil is a major carbon sink, an essential part of the climate change solution.

A greater understanding of our soil and its contribution to the stability of the planet is an important step in securing a healthy future for us all in the years to come. With an increasing world population and decreasing land-mass, food insecurity is an ever-more pressing concern. We thought that the answer was to pump fertilisers into plants at one end and wait for greater yields at the other. Now, we understand that the delicate relationship between soil and plants is more subtle than that. Soils can increase crop yields and address food insecurity, but we need to work with the natural processes inherent in their structure and biology.

PART TWO:
GETTING TO KNOW YOUR SOIL

SOIL IN OUR GARDENS

Each time I am drawn by aspirational photographs to visit a famous garden, I have to brace myself for an attack of soil envy.

DIFFERENT TYPES OF SOIL

Why is it that everyone seems to have 'better' soil than me? Their plants glow with vigour and seem to have no problems with malnutrition or disease.

In this chapter we will look at the differences between soils, why they vary and ask the question, 'Is one soil really better than another?'

You may have heard people talk about 'a clay soil', 'a chalky soil' or 'an acid soil'. These terms only describe one aspect of a soil's character – but a soil has many sides to its character and to fully understand your own soil, you need to be familiar with its whole personality. It's a bit like describing someone as 'shy' when there is a lot more to them than that if you take the trouble to get to know them.

Like a person, there will be things about your soil that you like and probably some that you are not so keen on. It is the balance of all the different parts that add up to an individual person or soil. In fact, your garden, balcony or windowsill is a unique growing environment. There is no other place on Earth that has that exact horticultural make-up, and much of it is determined by the soil.

When getting to know your soil, you will be observing what it looks like, delving into its chemical reactions, watching how it behaves with water, seeing which plants grow well in it, and getting your hands dirty as you feel its particles between your fingers. Even a good sniff will shed light on its health and composition.

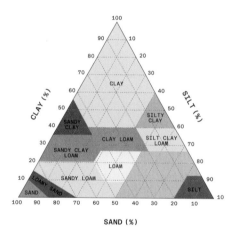

Above: This well-known diagram shows how soils can be a combination of different proportions of soil particles: silt, clay and sand.

Opposite: Landscapes are a complex tapestry of soil types laid down by different local processes over millennia.

SOILSCAPES FOR ENGLAND AND WALES

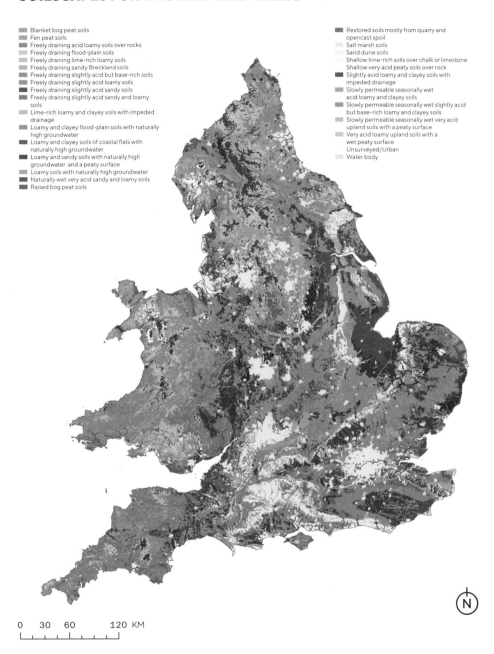

- Blanket bog peat soils
- Fen peat soils
- Freely draining acid loamy soils over rocks
- Freely draining flood-plain soils
- Freely draining lime-rich loamy soils
- Freely draining sandy Breckland soils
- Freely draining slightly acid but base-rich soils
- Freely draining slightly acid loamy soils
- Freely draining slightly acid sandy soils
- Freely draining slightly acid sandy and loamy soils
- Lime-rich loamy and clayey soils with impeded drainage
- Loamy and clayey flood-plain soils with naturally high groundwater
- Loamy and clayey soils of coastal flats with naturally high groundwater
- Loamy and sandy soils with naturally high groundwater and a peaty surface
- Loamy soils with naturally high groundwater
- Naturally wet very acid sandy and loamy soils
- Raised bog peat soils

- Restored soils mostly from quarry and opencast spoil
- Salt marsh soils
- Sand dune soils
- Shallow lime-rich soils over chalk or limestone
- Shallow very acid peaty soils over rock
- Slightly acid loamy and clayey soils with impeded drainage
- Slowly permeable seasonally wet acid loamy and clayey soils
- Slowly permeable seasonally wet slightly acid but base-rich loamy and clayey soils
- Slowly permeable seasonally wet very acid upland soils with a peaty surface
- Very acid loamy upland soils with a wet peaty surface
- Unsurveyed/Urban
- Water body

0   30   60   120 KM

N

In this following chapter, I will take you through some simple tests you can do at home to learn about your soil's make-up, but before that I will describe some of the most common soils.

Let's start with the main object of my soil envy - loamy soil. Generations of gardeners have sought to 'improve' their soil, striving to achieve a 'loamy consistency'. Traditionally we were taught to add sand, grit, lime, calcium or gypsum to clay soil in order to 'open up' its structure. Then we were told to add organic matter to sandy soil to help improve its structure. Going further, to try and change the chemical composition of our soil, we can add lime to make it more alkaline; sulphur to make it more acidic; and gypsum, calcium and a whole armoury of fertilisers that are recommended to improve our soil and make our gardens 'better'.

Let's take our foot off the fork and think for a moment. If you sit and look at your garden plot and then look over the fence to the surrounding countryside, it is worth asking ourselves: what does 'better' mean?

I think it means that we all want our gardens to look like the books. Yes, I want to visit a famous RHS garden and smile smugly to myself, 'Pretty much on a par with my rose border at home!'

Since the beginnings of the home gardening industry in the early twentieth century, we have been seduced by increasingly airbrushed images of impossibly perfect blooms, fruit and vegetables. We believe that our gardens should all look alike and that to have 'perfect' plants, we need to cultivate 'perfect' soil.

However, there are now important questions being asked about whether this is wise, desirable or even possible.

Why do the Yorkshire Moors look and feel different to the South Downs? It is largely due to the difference in their soil, derived from differences in the underlying rock, that has led to distinct flora and fauna, unique to each area. There are plants that thrive on the moors that would struggle on the downs and vice versa. Those plants give each area its local character – much celebrated in the landscape, but often forgotten when we come to private gardens.

Should we all be trying to create identical gardens and soils? Should a garden in Yorkshire look the same as a garden in Sussex? How great is the environmental cost of changing our soils and what do we lose in trying?

At this point, you might be wondering why I have written this book and why you are reading it! This book is about seeking a healthy soil, in the belief that the healthiest soil for your location is the best soil. It is not about changing your soil to another type, but more about avoiding damage, enhancing its environmental processes and working with the natural quirks of your site. And choosing the plants that are adapted to your area to achieve a healthy garden of local individuality.

Some people feel blessed with their loamy soil, but I want to encourage the rest of you not to feel cursed. So, here we go then: five of Britain's most common soils.

LOAMY

Loamy soil, or loam, has for many years been the Holy Grail of the gardening world. It is an ideal mix of the three different soil particle sizes – sand, silt and clay – and so is generally thought to be the best of all worlds. It is not too gritty and free-draining, and not too claggy and unworkable – the Goldilocks of soils.

From the diagram on page 40, you can see that loam sits at the middle of the soil texture triangle. However, if its composition changes, moving towards one or other point of that triangle, it takes on a different character. For example, if loam is a little more than a third clay, it becomes clay-loam; if it is a little more sand, it becomes sandy-loam and if it has a little more sand and clay, it becomes sandy-clay loam.

So, a balanced loam has enough large sand particles to allow air to be held in pockets between the particles, making it light and well aerated. This means that its structure is open and easy for plant roots to penetrate with plenty of available gases for their metabolic activities. There is also space for water to move freely between particles, making it available for use and preventing the soil from becoming waterlogged and sour.

On the other hand, the silt and clay particles are small enough to give loamy soil the ability to hold on to water well, so it does not dry out too rapidly, as pure sand does. And a soil that holds water is also holding the nutrients dissolved in the water, making them available to plant roots to take up.

CHALK

Chalk and limestone derive from the shells and bones of ancient creatures that fell to the bottom of seas aeons ago, and under the weight of time and geological processes have become deposits of calcium carbonate in the soil.

The calcium carbonate makes the soil alkaline and if it is present in chunks, you will see bits of white chalk visible in the soil. In chalky soils you may also see flints, and in limestone soil pieces of limestone.

The structure and texture of chalky soils is very variable, depending on the form that the chalk takes and whether there is also clay present. It can be heavy and clay-like, or it can be gritty, light, free-draining and dry. If a soil is free-draining, the important nutrients – potassium and nitrates – tend to dissolve and drain away as the water does, making the soil poorer.

If your soil is clay-like, it will hold moisture and more soluble nutrients. However, the chemical availability of the potassium will also depend on the type of clay soil, some are 'potassium-releasing', whereas others hold potassium ions more tightly within their structure.

Plants growing in chalky soils can be prone to iron deficiencies. This nutrient is essential for photosynthesis and a lack can be detected as yellowing of the leaves (lime-induced chlorosis), particularly in an angular pattern between leaf veins and in the youngest, faster growing parts of the plant. It is rarely caused by a lack of iron in the soil, but a high soil pH that prevents the plant from being able to take up the solid form of iron.

Plants vary in their susceptibility to iron deficiency, so if you have a chalky soil, it is best to avoid acid-loving plants.

CHALK GARDENS TO VISIT
- **Hidcote Manor**, Chipping Campden, Gloucestershire
- **Highdown**, near Worthing, West Sussex
- **University Botanic Garden**, Cambridge
- **Waddesdon Manor**, near Aylesbury, Buckinghamshire

CLAY

A soil is said to be clay soil when at least a quarter of its particles are clay. Clay particles are the smallest and have a plate-like shape, so they tend to slide over each other and have smaller and fewer air spaces between.

This gives clay soil its particular characteristics. You can tell you have a high clay content in your soil if it holds its shape when you roll it into a sausage with your fingers. It feels smooth and sticky, and when wet with clay, your fingers will slide over each other.

It is easy to imagine water trapped between the microscopic layers of particles, and with the water, nutrient ions become trapped. Therefore, clay soil tends to hold on to its water and its nutrients. The nutrients are in solution (dissolved in the water), so readily available to the roots of plants therein.

Clay soil is wet, heavy and claggy in wet weather, making it cold and difficult to dig. However, if clay soil dries out, it becomes difficult to re-wet – it 'bakes' hard in the sun and becomes unworkable in a different way.

The structure of clay soil is particularly delicate and easy to damage if you work it when it is too cold or wet. It can become compacted and puddle, losing its ability to drain. This can then cause waterlogging and airlessness. If this situation continues, at its worst soil microorganisms will die in the anaerobic conditions, causing the soil to become sour and smell of rotten eggs (sulphur). Eventually the dead soil will take on a grey or black appearance.

CLAY GARDENS TO VISIT

- **Hyde Hall**, Chelmsford, Essex
- **Rosemoor**, Torrington, Devon
- **Beth Chatto Gardens**, Colchester, Essex
- **Marchants Hardy Plants**, Laughton, East Sussex
- **Elizabeth Gaskill's House**, Manchester

SILT

Silt particles have a size between large sand and the smallest clay particles. Silt particles are formed from eroded rocks such as granite and are usually transported by water, ice or wind to their location. Silt carried by water can cause rivers or wetlands to 'silt' up, or become blocked with silt, and winds can create silt 'blankets' across inland landscapes.

A soil is called silt soil if it is composed of at least 80% silt particles, so this is a rarer type in the UK. Silts are usually nutritious soils, more so than clay and sand. Silts have good aeration and a structure that allows plant roots to penetrate and metabolise. They are able to hold on to water better than sands, helping plants to withstand drought for longer.

A silty soil will look and feel fine and slippery. It does not clump easily and has a delicate structure which is easy to compact and damage.

It is likely that your soil will have a silt component to it, combined with loam or clay.

SILTY GARDENS TO VISIT
- **St John's Jerusalem**, Sutton-at-Hone, Kent
- **Isabella Plantation**, Richmond Park, London
- **Beth Chatto Garden**, Elmstead, Essex
- **RHS Garden, Bridgewater**, Salford, Greater Manchester

SAND

Sand particles are the largest. They are easily seen with the naked eye and felt as gritty particles of pulverised rock.

You will know if you have a very sandy soil. It will be loose and dry because it does not hold water well, and will therefore also be low in nutrients because it cannot hold on to the nutrient ions between soil particles. Sand also warms up quickly and holds on to that warmth.

Most sandy soils will be a mixture, with some sand particles as well as loam or clay. However, you will be able to feel a slight grittiness if there is sand in there.

SANDY GARDENS TO VISIT

- **Derek Jarman's Garden**, Dungeness, Kent
- **Knoll Gardens**, Wimborne, Dorset
- **York Gate**, Leeds, West Yorkshire

HOW TO TEST YOUR SOIL

If you want to understand your soil, start by taking a walk around the neighbourhood; your best method is observation.

WHAT HEALTHY SOIL LOOKS LIKE

Have a look at the natural landscape first if you can. The plants that are growing naturally, perhaps on your local common or in the woods, will tell you a great deal. For example, if you live by the coast you will know that the soil is coastal sand, and if it is sandy heathland, you may see pine and birch trees with gorse, bracken and heather. Shallow chalk soils support grasslands rich in wild flowers such as orchids, knapweed and devil's bit scabious. Looking at other gardens can also be useful. For example, acers, rhododendrons and camellias indicate acid soil.

Next, have a look at what is growing in your own garden. It is possible that there are different things going on in different parts of the garden, so observe:

- How are the plants getting along in the different parts of the garden?
- Are there any tell-tale weeds, such as:
  - plantain, creeping buttercup and creeping thistle, which love heavy and compacted soil?
  - common groundsel and stinging nettles that indicate nutrient-rich soil?
- Most mosses indicate dampness
- Yarrow tolerates poor sandy soil

Now to the soil itself. The number one question is: does your soil look healthy?

Look first at the surface. Are there areas of compaction or waterlogging, which may be covered with a film of algal growth or moss? Or are there areas of excessive dryness?

When you hold a handful of soil and crumble it between your fingers, a healthy soil will have a variety of fine crumb sizes (aggregates) with large and small air pockets between. The ideal structure is like an apple crumble topping, with fewer large lumps of sugar! The more large clods of earth there are stuck together, the less good the soil structure.

More organic matter tends to make soil darker in colour (pale grey is generally not a good sign).

Healthy soil will feel moist and have a fresh, earthy smell. If it does not smell at all, it probably has low levels of life and organic matter.

Digging into your soil, if it is healthy, you should come across life, in particular earthworms. In 20 cubic centimetres/8 cubic inches of soil, it is considered good to have more than thirty-five worms and poor if you have fewer than fifteen.

PLANT YOUR PANTS!

When assessing soil life, scientists observe it under a microscope, or use a soil microbiometer (a test for microbes in soil). These methods require expertise, equipment and expense.

One very simple and fun test to try is to plant a pair of organic cotton pants (or any other 100% cotton item of fabric or clothing). If your soil is biologically active, the pants will have pretty much deteriorated after eight weeks. It is a great way to engage children with soil and also a good comparative test for different soils.

www.countrytrust.org.uk/plantyourpants

Below: A very visual and memorable way to demonstrate how organisms in a healthy soil will rapidly degrade organic material.

HOW TO TEST YOUR SOIL'S PHYSICAL CHARACTERISTICS

Your senses are a great tool for an initial assessment of your soil's health and composition.

Looking first, its colour will give you a clue. Dark, brown, crumbly soil is usually rich in organic matter; generally, the paler a soil the less organic matter it has. You may also be able to see chinks of white chalk in a chalky soil, grains of sand or clods of thick clay.

When you smell it, you might smell a sweet earthiness if it is rich in organic matter. However, if it is unhealthy and anaerobic, it may smell sulphurous and sour.

Next, feel your soil between your fingers and squeeze it in the palm of your hand.

If you happen to have a 'picophone', you could even listen to your soil. Scientists from the University of Zurich have found that the more organic matter and life there is in a sample of soil, the noisier it is, with all the organisms coming and going.

1.
Choose a day when there has been recent rain, so the soil is moist but not sodden.

2.
Remove the plant material from the surface of the soil and dig a handful of topsoil to feel.

3.
Do you see any lumps of chalk, flint or stones?

4.
Squeeze the soil in the palm of your hand to see if it clumps together – this would indicate a significant proportion of clay.

5.
Feel the soil between thumb and forefinger – if you can feel grittiness, then it has a significant amount of sand.

Right: Feeling, crumbling and squeezing a handful of soil may give a good indication of whether it is chalky, sandy or has a high clay content.

SOIL SEDIMENTATION TEST TO DETERMINE PARTICLE PROPORTIONS IN YOUR SOIL

If you want more specific information about the proportions of sand, silt and clay in a soil sample, you can use the soil sedimentation test.

Because sand, silt and clay have different sized particles of different weights, they will separate out at different speeds if mixed with water.

If you mix a soil sample thoroughly with water and time how it settles again, the speed of deposition and the layers that are formed will reveal the soil composition.

Start with a tall empty jar, a marker pen and some water.

1.
Take a sample of dry soil; remove any stones and break up any clods.

2.
Fill a jar one-third full with the dry soil.

3.
Fill the jar with distilled water to about 3cm/1¼in below the top.

4.
Place the lid on the jar and shake it vigorously to mix the soil and water well.

5.
Place the jar on a level surface, and wait for the particles to settle.

Results and Interpretation

Sand
Mark a line after 1–2 minutes.

Silt
Mark a line after 1 hour.

Clay
Mark a line after 24 hours, or when the water is relatively clear.

SOIL PROFILE

A complete, formal soil profile undertaken by soil scientists in the field is quite an undertaking in terms of time, equipment and analysis. They are required to dig a pit approximately 1m/39in wide and deep in order to observe all the layers of soil formation.

For your garden a smaller pit size and depth is still likely to yield useful information in terms of the depth of organic matter, topsoil and the type of underlying subsoil.

Some soils have a very complex make-up of soil layers or horizons, all of which have a distinct colour and texture, to enable you to easily identify them. For other soils, perhaps only some of the horizons are visible. Some may be very thin and others so deep that you cannot get to the bottom.

Here I have outlined the four main horizons which are commonly present and easy to see.

The organic layer, topsoil, subsoil and bedrock can be clearly interpreted in terms of the weathering of soil from rock and its development from organic matter.

1.
Dig a straight-sided pit in the soil about two spades deep (you can go deeper if this depth does not give you the information you need).

2.
Observe the horizons of the soil profile and measure the depth of each soil layer:

- Organic layer (O horizon) – the only layer not dominated by minerals, present in different thicknesses or not at all.
- Topsoil (A horizon) – mainly mineral layer, but with high organic content. Darkest layer in colour, containing the most soil life.
- Subsoil (B horizon) – often mostly material washed in from the layers above.
- Parent bedrock (R horizon).

Right: Although a soil may have more layers, the organic layer, topsoil, subsoil and bedrock are often clearly visible.

SOIL HORIZONS

O — ORGANIC LAYER
HUMUS

A — TOPSOIL
MINERALS WITH HUMUS

B — SUBSOIL
DEPOSITED MINERAL
& METAL SALTS

R — BEDROCK
UNWEATHERED PARENT ROCK

SOIL PERCOLATION AND DRAINAGE

Soil permeability affects how much a soil holds on to water and therefore also nutrients. It also affects how well aerated the soil can be and how easily roots can grow through that soil. The permeability of a soil can be affected by:

- Soil texture
- Soil structure and how large the aggregates of particles are
- How large the pores between the aggregates are
- The amount of organic matter, which may open up the structure of the soil
- Moisture content
- Soil compaction, which may impede water movement

There are complex scientific tests for soil permeability that require specialist equipment; however, you can easily do this simple test in your garden and it will give you a good idea of how well draining your soil is.

1.
Dig a hole 30cm/12in wide and deep (approximately one spade blade length size).

2.
Fill it with water (approx. 20 litres/5 gallons) and leave it to drain overnight (this saturates the soil).

3.
Refill it with water and measure the depth it drops to every hour.

4.
Calculate your soil's drainage rate (height change in cm/in per hour).

5.
Interpret your results using the table below.

Drainage Rates and Interpretation

<2.5cm/1in per hour
Drainage is too slow

2.5–7.5cm/1–3in per hour
Good drainage

>10cm/4in per hour
Drainage is too fast

Opposite: Filling a pit with water will give you a good idea of how easily rainwater will seep through your soil.

SOIL AGGREGATION (SLAKE TEST)

A well-structured soil has small clumps of soil particles with air and water in spaces between the aggregates. In a healthy soil, the aggregates of particles are held firmly enough together for the soil to absorb water without the structure crumbling apart.

In a slake test, you gently submerge a clump of soil into water and time how long it takes to disintegrate.

1.
Choose a clump of soil to be tested from a sample of air-dried soil.

2.
Fill a jar with water.

3.
Place a piece of mesh into a small beaker.

4.
Place the clump of dry soil on to the mesh, fully submerged in the water.

5.
With a stopwatch time how quickly it takes for the clump to disintegrate.

Results and Interpretation

Poor soil aggregation
The clump of soil disintegrates and falls apart in less than 2 minutes.

Moderate soil aggregation
The clump of soil disintegrates and falls apart in 2–10 minutes / a small portion of the clump remains intact.

Good soil aggregation
The clump of soil disintegrates and falls apart in more than 10 minutes / a large portion of the clump remains intact.

This test is good for comparing different parts of your garden or your soil before and after treatment.

HOW TO TEST YOUR SOIL'S CHEMICAL CHARACTERISTICS

It is a good idea to test your soil in a few different places around the garden, because over the years additives may, in theory, have altered the pH of your soil. Also in some areas, soil pH can vary across a site.

If you are only able to test at one position, the best place is the soil under your lawn, because it is unlikely that anyone would have altered the pH there.

You can use a simple chemical test kit from the garden centre (I do not recommend using a cheap pH metre).

1.
Dig a trowel blade's depth into the topsoil and remove a sample of soil.

2.
Clear out any stones or organic matter.

3.
Crumble a small sample of soil into the bottom of your test tube.

4.
Add enough of the test reagent to the marker on the side of the test tube.

5.
Put the lid on and shake your test tube vigorously to dissolve the soil in the reagent.

6.
Leave the sample for the soil to settle and the colour to become clear (this may take some time with a fine clay soil).

7.
Once the colour change in the liquid is apparent above the layer of dissolved soil, compare the colour with the colour chart.

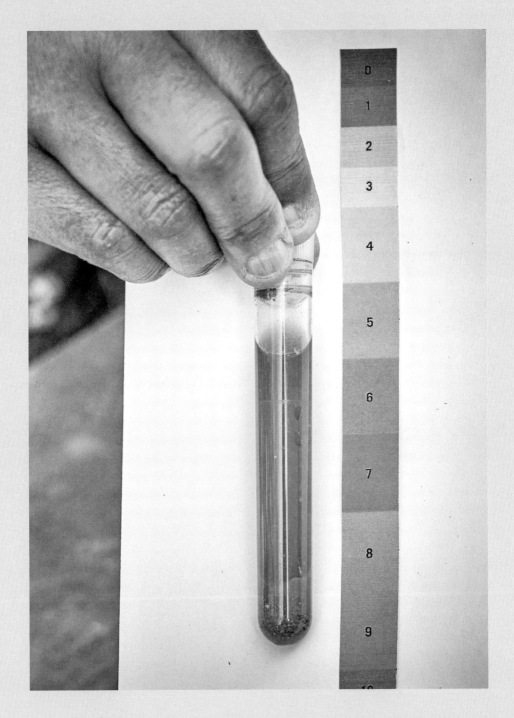

SOIL THROUGH THE YEAR

Soil is a dynamic environment with varied constituents of water, air, minerals and organic matter including micro and macroorganisms.

These are constantly undergoing chemical, physical and biological processes. There is a lot going on in your garden soil and it is all affected by internal and external environmental conditions.

Soil acidity, potassium, nitrogen and phosphorus fluctuate with the seasons, but generally these changes are too small for a home gardener to measure or be concerned with.

However, there are some seasonal changes in your soil which you do need to be aware of. Changes in rainfall, temperature, winds and biological life cycles will alter the processes, content and composition of your soil and this might well affect your soil management and gardening practices.

When there is heavy rainfall, it can cause erosion of the soil and leaching of soluble nutrients away from the plant root zone. Cover crops, winter weeds and mulching will all help to protect the soil.

Wet soil, particularly if it is also cold, is prone to compaction and physical damage when dug over. The best time to work delicate soils, like clay and silt, is in the autumn, when the soil is still warm after summer but the winter rains have not started. If you turn the soil in the autumn and leave clods exposed, the repeated freezing and thawing over the winter months will help to break up the clods and make it more workable the next year.

When the soil cools and reaches about 7°C/44°F, biological activity slows. Larger creatures, like earthworms, and microorganisms will slow down and the greenery of most plants will die down.

As the soil warms again, the biological processes start up and a new season of growth begins, along with the accompanying chemical and physical changes in the soil.

At this time, when the soil warms, light, sandy soil can be cultivated. Later in the summer, light soils will be prone to wind erosion and clay soils may dry out to the extent that they cannot be dug at all.

So, it helps to plan your garden management and work with the seasons and weather variations.

Right: Under a blanket of snow, many organisms undergo a slowing of their biological processes until the soil warms and the growing season begins again.

SOIL IN OUR HOMES

To thrive in the alien environment of our homes, plants need us to step in with some extra TLC.

HOUSE PLANTS

The soil environment in your garden is complex and extensive. Animals, minibeasts and microorganisms are constantly moving in and out and up and down the soil layers. Water and chemicals flow around and biological processes take place. None of these players has any consideration for the imaginary boundaries that we might invent: garden fences, borders and land rights. Their world is vast and only restricted by our interventions to create physical barriers.

One intervention would be to put soil, along with its inhabitants, into a pot. Depending on the size of the pot, they can continue happily for some time, but eventually this artificial containment will take its toll on their natural processes. A plant container is a highly abnormal place for soil organisms and plants to live, so they need quite a bit of help to survive.

Now imagine taking that pot inside a house. Our homes are generally hot, dry, sealed and environmentally constant. No natural rain, no changing of temperature through the day and seasons. Even the light will go on and off at strange times.

House plants need nurturing – they are often 'foreign' plants in a foreign environment. The antithesis to the garden mantra, 'Right Plant, Right Place'.

The secret of success is to really understand the natural habitat and requirements of the plant that you want to grow and then give it as close a version as you can. That starts with the soil.

Above: A pot of soil is a confined and abnormal environment for soil organisms and plants to survive.

COMPOSTS FOR HOUSE PLANTS

The starting point for healthy potted plants is a good quality multipurpose, peat-free compost. Most plants will be okay with this, but taking a little time to create a mix tailored to each plant's requirements will help them to thrive.

Most house plants have thin fibrous roots, so creating a mix that has air spaces and is fairly free-draining is ideal. A mix of 50% peat-free compost, 25% horticultural grit and 25% fine bark is a good general-purpose house plant mix.

If you are growing cacti, they hate water sitting around their roots and stems, so make the compost even more free draining by increasing the grit component to 50%, combined with 50% peat-free compost.

Fleshy rooted plants like the Swiss Cheese plant prefer an open but moisture-retentive soil, so use 60% peat-free with 20% fine bark and 20% coarse bark.

Orchids have fleshy roots, but also hate to sit in water. They come from humid climes, and many are epiphytic – growing naturally without any soil. Most orchid compost recipes will start with a mix of barks of different sizes, depending on the root size of the plant. This anchors the plant whilst allowing air and water to circulate. Other ingredients may include coir, perlite and moss. It can get complicated to provide the right conditions, along with the correct nutrient balance, so unless you are growing a number of orchids, it may be easier to purchase a mix from the garden centre.

Below: Composts for house plants are an artificial mix of ingredients to provide the necessary aeration, moisture, drainage and nutrients.

SOILLESS GARDENING

In a book all about the importance of soil, you may be surprised to read that many plants can grow happily without any soil at all!

AIR PLANTS

An epiphytic plant, of which orchids are one example, is a plant that has evolved to grow on another larger plant, without soil. They tend to live perched up high, where they can get enough sunlight and also benefit from the nutrients in rotting leaves and debris falling from the plants around them. They absorb moisture from the atmosphere and often minerals dissolved in rainwater.

When we grow epiphytes in our homes, they need physical support – it is fun to get creative with where you are going to put them! However, our houses are usually devoid of rotting organic matter and nutrient-rich rainfall, so we need to provide the necessary supplements for an epiphyte's survival. Fortunately, they do not mind being potted, as long as the mix is airy and does not get waterlogged. A bark-rich mix with moss and perlite and perhaps a little soil or peat-free compost is ideal.

Please note: any sphagnum moss should only be purchased from a reputable farm source and not taken from the wild.

HYDROPONICS

I first looked into hydroponics when I was researching ways to grow food in a small, urban home, without garden space. These are artificial systems that involve growing the plants in water, without soil, and so they are clean and potentially space-saving method.

The idea is that you provide the roots of the plant with all the water and nutrients that it needs, directly, without using soil as a delivery medium. The plants roots are suspended in a nutrient bath, which circulates continuously. The nutrient levels are carefully controlled and adjusted according to the life-stage and requirements of each plant

Leaving out the soil means that you also avoid problems with soil-borne pests and diseases. The plants are bathed in nutrients, so there is no need for the roots to search out what they need and expend energy pushing through resistant soil layers. It is an indoor system, so with controlled artificial lighting you can extend the growing season – which would more often be restricted to the summer months – to twelve months of the year.

Most plants can be grown hydroponically, but it is most often used for salad crops, strawberries and some cut flowers.

The composition of the nutrient bath can be fine-tuned to the specific requirements of each plant and its stage of growth. The plants need physical support and this can be a solid substance or suspension system.

If you add fish to your growing system, it is called aquaponics. The fish create nitrogen-rich waste, which can be recycled into the growing areas to fertilise the plants. Consider that at this point it can get quite technical, because you are supporting fish life as well as plant life.

RUBBLE AND SAND GARDENS

I live near the South Downs in the south of England and here we have protected chalk downlands of designated world importance. On the shallow, nutrient-poor soils of the Sussex hills around my home, there are myriads of wild flowers that thrive with little competition from nutrient-hungry grasses. Upon these wild flowers further myriads of insects feed, and a whole, complex ecosystem is sustained. It is the nutrient poverty of the soil which is key to this particular ecosystem in this part of the world, and in many others.

As we open our eyes to look more carefully at the natural world and open our minds to the importance of every species and every ecosystem, we start to understand that even poor soil, derelict land and piles of neglected builders' rubble can be home to many species.

As well as considering our own gardens, we need to preserve and even create rubble, sand and brownfield gardens in order to restore biodiversity. There is a growing trend for these gardens and it is important that we do not start spreading rubble and sand across the land willy-nilly. As always, assessment of the existing site is key and respect for the species and soil therein. Rubble gardens may well be a suitable solution for a brownfield urban site and a sand garden may be the quickest and most appropriate if that is the soil that you have.

It is often drought-tolerant plants that appreciate the free-draining, poor soils of chalk, sand and rubble most. Some of these habitats and the plants they support will become increasingly important as we navigate a warmer climatic future.

RUBBLE AND SAND GARDENS TO VISIT

- **Knepp Castle Walled Garden**
  Horsham, West Sussex
- **Brownfield Garden**
  Milton Park , Oxfordshire
- **The Sand Bank**
  Beth Chatto Gardens,
  Colchester, Essex
- **The Beth Chatto Community Garden**
  Colchester (temporary, reversible, urban brownfield garden)

Previous page: Organic vegetables grown using hydroponic farming.

Right: The sand garden at Knepp Castle was previously an English-style garden where traditional soil improvement would have been de rigueur.

PART THREE: GARDENING FOR BETTER SOIL

COMPOST AND MULCH

People use the confusing term 'compost' interchangeably for a range of different things, and there is also a host of compost products available.

DIFFERENT TYPES OF COMPOST

Simply put, compost is any decomposed organic matter, but that definition does not help us with distinguishing one type from another.

Let's start by separating the compost you can make at home, at almost no cost, from those composts produced in bulk to buy from the garden centre or plant nursery.

COMPOSTS MADE AT HOME
Garden compost

The most common compost that you can make for yourself is garden compost from rotting down garden and kitchen waste.

Vermicompost

This is the compost that worms will make for you if you feed them vegetable scraps in an enclosed container.

Leaf mould

Leaf mould is also a type of compost; however, we tend not to refer to it in that way. It is most commonly used as a mulch, so I have written more about it on page 82.

Chicken manure

If you keep chickens at home, their soiled bedding is a great addition to the compost heap. The urine will activate the rotting process of the other ingredients and is rich in nitrogen.

BOUGHT, BAGGED COMPOSTS

Commercial composts can be bought in small bags from a garden centre, where the choice is dizzying. If you need large quantities, you can buy very large bulk bags, or even get supplies tipped loose on to your driveway. Buying in bulk is cost effective, but the choice of compost varieties is more limited.

These composts can be further divided into those that are harvested from natural sources and those that are manufactured. In the second instance ingredients are mixed in specific quantities and the composting process activated.

As well as varying in composition, composts also vary in how biologically active and 'alive' they are, because some are heat-treated, which kills organisms. Measures of the biological activity of different commercially available composts are not generally available.

Right: Different composts from kitchen scraps, worm composting, leaf mould and animal manure.

NATURALLY PRODUCED AND HARVESTED COMPOSTS

Loam

Loam is a compost that is rich in topsoil. Traditionally it would have been a by-product of the turf industry, the soil shaken off after cutting and stacking newly collected turves (sections of lawn).

Alternatively, areas are stripped of their topsoil, which is sold as loam.

These days loam is also a by-product of growing sugar beet for cattle or sugar for people.

Manure

Animal manure from pig or chicken farms, farmyards and stables can be collected and used after being stored for a period of time or heat treated. This is to ensure that it is well rotted; otherwise it may scorch your plants.

There are a couple of potential problems with manure: if the animals have been roaming free in the fields and eating weeds, the weed seeds can pass through the animal unscathed and germinate in your planting bed. Storing and allowing manure to heat up during the rotting process will hopefully kill the weed seeds.

Another less common problem is that if the animals have been fed plants that have been sprayed with herbicides, the chemicals can survive in the manure and stunt the growth of your plants. Unexplained slow growth, with pale, yellow, wizened leaves, is a sign of this.

ARTIFICIALLY FORMULATED COMPOSTS

John Innes composts

JI composts originated in the 1930s at the John Innes Horticultural Institute then in Surrey. They contained loam, which gave weight and water-holding properties, making them superior to other loam-free mixes. These days loam is in short supply, so other ingredients may replace loam in JI composts.

John Innes No. 1 – formulated for seedlings and young plants.

John Innes No. 2 – formulated for hungry house plants or vegetable plants in pots. It has twice the nutrients of JI No. 1.

John Innes No. 3 – this is the most nutritious formulation, with three times the nutrients of No. 1. It is formulated for mature plants in containers or hungry vegetables, such as tomatoes.

All-purpose

All-purpose or multipurpose compost, as the name suggests, is suitable for many applications. Until recently, it would have had a peat base, but home peat use has now been banned in many countries, including the UK. For many years manufacturers have been trialling peat substitutes and there are now several alternatives available; most of these do not hold water as well as peat, so watch out for potential drying out.

Seed

A fine-textured mix, formulated for seed sowing.

Cuttings
A fine-textured and nutritious mix for young seedlings and cuttings.

Tree, shrub and rose
A heavier mix with plenty of nutrients to sustain larger, mature plants in pots and hungry vegetables, such as tomatoes.

Ericaceous
Acid-loving plants such as acers, rhododendrons and camellias used to be grown with peat-based composts, which have the low pH that they need.

Now that peat is unavailable for domestic use, there are peat-free alternatives that have been formulated to support these plants.

Below: There is a dizzying array of bagged composts to choose from.

INGREDIENTS OF PEAT-FREE COMPOST FORMULATIONS

Coir

A by-product of coconut production, coir is the shredded, fibrous outer layers of the coconut. It holds water well and so is used to replace peat. It is imported into the UK from the Indian subcontinent.

Wood fibre

This by-product of the timber industry is expanded wood chips. If left to rot down naturally, that process can leach nitrogen out of the soil (an undesirable effect), so some manufacturers add nitrogen to the wood chips to counteract the detrimental effect.

Composted bark

Another by-product of the timber industry, the small pieces of bark create air pockets and improve drainage in a compost. They too can leach nitrogen as they decompose, so some of these products may also have added nitrogen.

Wool composts and bracken composts

These formulations use waste and weed products, such as composted sheep's wool or bracken (ferns) as the base for the compost.

Council green garden compost

This is similar to your own homemade garden compost, but is created in bulk by your local council from household collections. Its advantages are that it saves space for householders; it is part of a circular local economy; it is generally fairly cheap; it is generally rich in nutrients, bacteria and fungi. However, the quality and make-up are variable, and unwanted detritus such as glass and plastic are often found within. Take care not to put such things in your own council green waste bins.

Spent mushroom compost

This mushroom farm by-product used to be leftover peat-based compost, so fibrous, moisture-retentive and nutritious. It was also usually alkaline due to chemical action of the mushrooms.

Peat is currently permitted in the growing of mushrooms, but will be set to change in the near future.

Specialist composts

Some plants have specific needs for healthy growth, so there are various formulations, usually in smaller bags, that are available to pamper your plants:

- House plant
- Orchid
- Cactus and succulent
- Mediterranean and citrus

HOW TO MAKE GARDEN COMPOST

The principles of composting

Decomposition is the natural process of rotting down that will happen in your garden whatever you do. It is the process by which any plant (or animal) material eventually becomes part of the organic matter in soil.

Composting is a way of harnessing this natural process in order to turn collected waste materials from your household into a useful additive to improve your garden soil.

Decomposition or composting involves various chemical reactions, which require organic matter (containing nitrogen and carbon), soil organisms, water, air and heat. The organisms, water and air act on the organic matter, in the presence of heat, to break it down into a dark, crumbly, nutrient-rich organic material – compost.

So, whatever type of compost container you choose, you will need to make sure that your system has enough waste material, organisms, air, water and heat.

There are various methods and exceptions to this standard aerobic composting, which I outline below.

Gardeners have always home-composted. However, in recent years a wider community of householders has learnt of its benefits and there has been a boom in research and availability of convenient home-composting methods.

Different methods for composting

You have a wide choice of composting methods to choose from, depending on what is most convenient for your living situation, your lifestyle and the time you want to spend on turning waste into 'black gold'.

Passive versus active composting

The absolute easiest way to compost is just to leave your garden waste in a pile in a corner of the garden. Eventually it will rot down, even if you do nothing, but being a passive method, it will take a year, or two.

If you want to speed things up, you will need one of the active methods below. They involve managing your compost to optimise and accelerate the rotting process.

In-ground or surface composting

Usually people want to make compost that they can spread around the garden where it is needed (surface composting). However, if you have limited space or limited ability to carry heavy compost from A to B, you might prefer 'in-ground' composting. This is a way of making your compost exactly where you need the soil improved and the nutrients applied.

The easiest in-ground method is to dig a trench or hole where you plan to plant a few weeks or months later. Trench composting has been used by pea and bean growers for eons. These plants love the extra moisture captured by the decomposing organic matter and the slow release of nutrients over the weeks that the plants grow. If, after harvesting, you cut the tops off the dying plants and leave their roots, complete with their nitrogen-fixing nodules, in the soil, it will be a fabulous environment for growing a different crop next season.

One example of a commercial in-ground composter is the 'Green Cone'. These are great for a vegetable patch or raised bed. The organic matter breaks down and the nutrients leach out of the container into the nearby soil.

You can also buy in-ground wormeries. These are designed to keep the worms in place whilst allowing their goodness to seep out into the surrounding soil.

If you use a Bokashi fermentation bin, bury the acidic pre-compost that it produces in a shallow trench to complete its composting cycle. Do not put it in direct contact with delicate roots of plants, which can be 'burned' by the acid (see overleaf).

Aerobic versus anaerobic composting

Most processes for decomposition of organic material involve biological and chemical reactions that require oxygen, known as 'aerobic'. All the methods described below are aerobic except for Bokashi fermentation.

Bokashi bin fermentation

The Bokashi method is different from aerobic composting in that it is an anaerobic fermentation process that requires yeast and bacteria, and takes place in a sealed unit at room temperature. You do not stir or turn it, as you would compost, because that introduces air; instead, you add your kitchen scraps in layers and compress them with a layer of 'Bokashi bran' to squeeze out the air pockets.

Bokashi bran is an absorbent medium which contains the required bacteria and yeasts to get the process going.

The Bokashi method produces an acidic pre-compost after a couple of weeks. This still looks like the food waste that you put into the bin, but it has been pickled and so is ready for the next step. Pre-compost is acidic, so cannot be added directly to plants or planting areas; it needs further processing for about a month. To do this you can either add it to your compost heap, bury it in a trench or bury it in a part of the garden that you designate as your 'soil farm'. This is simply an empty corner of the garden where you condition the soil with Bokashi, and then harvest the improved soil from there to spread wherever you need it.

A Bokashi bin will also produce a leachate (liquid), which you need to regularly draw off via the tap, so that the contents of your bin don't get too soggy. The leachate of a Bokashi bin is different to compost tea; it is acidic and should not be used to fertilise plants directly. Instead, use it to improve infertile soil enhance your compost pile or dilute it to at least 1:100 to fertilise plants, indoor and out.

Bokashi bins are a convenient, clean and easy way of processing kitchen scraps, including meat and dairy. They are not suitable for all situations, but can be a very useful part of a wider composting system.

Cold versus hot composting

The process of composting requires heat, but it also produces its own heat. So, if you have a compost pile in a corner of your garden, it will produce enough heat on its own to slowly rot down. However, it is unlikely to reach high enough temperatures to kill off soil pathogens or unwanted weed seeds. So, you need to be extra vigilant about putting any detrimental waste into your pile - removing cuttings from sick plants or weed seed heads to your general waste instead.

Therefore, there is a lot to be said for working to retain the heat of your compost. Depending how much work you put in, your compost could be warm or hot. Hot composters operate at 40–60°C/104–140°F, making them very quick at decomposing waste into a finished product. They are basically very well-insulated compost bins designed to trap the heat produced and thus accelerate the rotting process.

Macro- versus micro-composting

The natural processes of decomposition generally involve large, macro soil organisms like woodlice and worms in breaking down raw material. Fungi aid the decomposition and microorganisms, including bacteria, finish off the process.

The systems described here involve macroorganisms to the greater or lesser extent that they find their way into the compost area. If you dig through your compost pile or watch it churning in your garden tumbler, you will see creepy crawlies rollicking in their element and,

importantly, hopefully you will see plenty of 'red wrigglers'. These are some of the composting worms that live in the surface layers of soil. They mature in a matter of weeks, laying eggs every ten days and eating prodigious amounts to produce similar quantities of phosphorous-rich 'poop'.

One way of accelerating the composting process is to 'farm' the worms. With 'vermiculture' you keep a community of composting worms in a suitably designed wormery and feed them waste. They then repay you with not only a fast turnover of compost, but also a rich liquid leachate, which you can collect and use as garden fertiliser.

A choice of home composters, food digesters and fermentation

- Trench composting
- Pile composting
- Bay composting
- 'Dalek' composting
- Rotating composting
- Hot composting
- In-ground worm composting
- In-ground food waste digesting
- Worm composting (vermiculture)
- Bokashi fermentation
- Compost club
- Council green bins

Above: Worms eat organic matter at an amazing rate. They pass it through their bodies, turning it into fine organic 'vermicompost'.

<u>Opposite and above</u>: There
is a composting method to
suit everyone, no matter
how much space you have,
how much you want to spend
or how squeamish you are.

HOW TO MAKE LEAF MULCH

Have you ever kicked the crisp autumn leaves of a beech tree into the air and noticed the dark spongy layer beneath? Decomposed beech leaves are gardeners' gold: rich, textured organic matter which is perfect for mulching planting beds and improving the soil.

Over many years these leaves, rich and fibrous, fall to the ground beneath the trees and rot into the earth, increasing the organic content. You can reproduce that process in your own garden, by creating a leaf mould cage and then spreading the contents on to your flower beds.

Please note: this leaf mould material is not strictly compost, because a pile of mainly leaves is broken down by fungi rather than bacteria. In a compost heap, you have a mix of woody material *and* green, sappy material and this is decomposed by bacteria.

1.
Mark out an area that is in shade in summer, but open to some rain. Approx. 1 × 1m/39 × 39in.

2.
Cover the area with a biodegradable weed suppressant layer, such as cardboard.

3.
Drive 75mm (3in) diameter tree stakes into the ground at each corner.

4.
Cut a length of chicken wire and secure it to each post with cable ties.

5.
Collect leaves that are thin, such as beech, oak, hornbeam and birch. If you have thick leaves, it is best to shred them. (Evergreen leaves are better composted.)

6.
Moisten a little in dry weather. Your leaf mould will be ready in 2–3 years.

Note: if you have restricted space you can make leaf mould by placing collected leaves in plastic bin bags. Pierce holes in the bags, moisten with water, seal and leave for a few years.

THE CASE AGAINST IMPROVING YOUR SOIL AND FEEDING YOUR PLANTS

Recent advances in science have resulted in challenges to our traditional ways of thinking about and treating soil.

SOIL IMPROVEMENT – TRADITIONAL WISDOM

I still have my mother's old *Reader's Digest Illustrated Guide to Gardening*. She and I would faithfully follow the steps to nurture our plants and improve our garden, including its soil. It has long been believed

that you need to work hard to improve your soil. Whatever you start out with, chalk, sand or clay, you need to try to turn it into the gardener's 'nirvana': a moisture-retentive loam.

That is because a moisture-retentive loam supports the widest variety of cultivated plants. The flowers, fruit and vegetables that we see in books have been bred to thrive in a generic soil and we all aspire to have those gardens that we see in the magazines.

The traditional methods I grew up with were, we now know, quite destructive and often counterproductive. Double-digging is not only boring and back-breaking, but destroys the structure of the soil, which has taken years to evolve. Remember the soil horizons we talked about earlier? Gone in an afternoon of double-digging!

Hoeing is less destructive; it tickles the surface of the soil to dislodge young weeds as they emerge. However, in doing that, you are also bringing thousands of dormant weed seeds to the surface, just waiting to come into the light and spring into action.

The rationale of 'no-dig' gardening is to neither disrupt the soil structure nor create an unending cycle of weeding. However, no-dig is still traditional in that it is about soil improvement. That is understandable; if your aim is to grow lush, healthy garden plants, fair enough.

I have concentrated in this book on garden soils that are nutrient rich for growing food and traditional ornamental gardens. However, many species, particularly natives, would not be happy in the pampered conditions of the average UK home gardener.

If you do have a garden with poor, shallow or free-draining sandy, gravel or chalk soil, you may want to improve areas so that you can grow food and traditional ornamental plants. However, a garden planted on unimproved soil and populated with species adapted to that soil, whilst not looking as showy, will have greater local character and distinctiveness. It will also be much easier to maintain, because you are working with the natural tendencies of the site.

FEEDING YOUR PLANTS – TRADITIONAL WISDOM

With traditional ornamental and kitchen gardening, at certain times of year we feed or fertilise the plants in order to encourage, first, leaf growth and then, later in the season, flowering and fruiting. This would be part of a gardener's yearly programme, usually with little thought about whether the plants actually need extra nutrients. We have been conditioned to believe that the more goodies we stuff into a plant, the more goodies we get out.

As I learn about soil, plants and the environment as a whole, I see the beautifully complex and balanced nature

Opposite: Traditional deep digging disrupts soil structure and processes.

Above: Feeding plants can be unnecessary or even detrimental.

of their systems and processes. When we disrupt these processes, we often do not have a full enough picture to anticipate all the consequences. We now know that when we add mineral nutrients to a soil, we may cause imbalances in the existing chemical structure, possibly switching off the very mechanisms that promote health in the plants we want to grow.

There are some occasions when you will want to fertilise your plants, but these are much less often than you might imagine. If you have a healthy soil rich in organic matter, or you are returning nutrients to the soil in the form of cuttings or manure, then the soil will be able to hold on to nutrients and make them available to plant roots. Extra chemicals will not be necessary. And if you are growing plants that are local, adapted to your site's conditions and suitable for the soil in its natural unimproved state, then they will thrive without extra minerals.

LEAN AND MEAN

Over the centuries of ornamental and productive gardening in Britain, we have sought ever bigger and showier plants through invasive cultivation techniques and increasingly complex plant engineering methods. As with animal breeding, often the results are genetically weak, prone to pests, diseases and lacking environmental resilience. To maintain these plants, we need to support and baby them.

Gardeners often complain that weeds don't seem to need any care; they just spring up healthily everywhere! That is because they grow where they are suited, no one looks after them and the fittest survive. Is there a lesson to be learned from nature here?

We can get away with being a bit meaner to our plants. Yes, they will be leaner and less showy. However, they will also be tougher and more resilient to the assaults of the weather and an increasingly unpredictable climate. In 1992, Beth Chatto created her Gravel Garden. She did improve the soil with organic matter, but having done so, she planted, mulched with a thick layer of gravel and, after initial establishment of the plants, never watered again. Plants that died were not replaced and so thirty years later, we have a living laboratory of drought-tolerant plants.

If you visit that garden, you will see that it does not look like a typical 'English Country Garden'; it has its own aesthetic, which reflects the character of the region, the climate and the story of the land. It is unique, as every piece of land is.

Careful observation of the nooks and crannies in your garden will reveal a variety of microclimates. These are pockets in the garden that have their own particular conditions, different to the wider space: for example, a wind tunnel down the narrow passage at the side of the house; a damp patch at the bottom of a slope where the rainwater runs to; an arid bed against the wall of the house; or a frost pocket, where nothing seems to survive the winter. Getting to know the complexities of your garden site will help you to choose the right plants for each part.

It is also informative, to watch what happens in your garden as the weather passes through and also as the climate changes over the years. Our weather has been increasingly variable and unpredictable in the last decade, with records being smashed with alarming regularity. Talk in the United Kingdom is of wetter winters and droughts in summer.

This is a deadly combination, because we cannot simply change to planting drought-tolerant, Mediterranean plants, as they may well drown in the UK winter. We need to look to plants that are resilient to different conditions, or perhaps to create growing conditions that support such resilience. There are plants adapted to every site in the world and your garden, although it may not feel like it at times, is no exception.

Like people, plants grown in healthy conditions tend to be more resilient. Good soil care and some organic matter will go a long way towards supporting general plant health, but overfeeding, without consideration, will make them sappy and weak. Some plants, particularly those adapted to dry conditions, will actually thrive better on poor soil.

MAKE YOUR OWN FERTILISER

WEED TEA OR NETTLE TEA/ COMFREY TEA

Many of your weeds can go on to your compost heap.

However, strong-rooted perennials (plants that live for more than one or two years), such ground elder and nettles, may take root in the compost or survive the composting process and later spread into your planting beds. These are best rotted down in water to make compost tea.

You can do the same with comfrey, a plant grown specifically for the purpose.

Both comfrey and nettles have high levels of nitrogen, potassium and phosphate. They are great for your tomatoes, roses and other hungry vegetable crops.

Harvest and chop up comfrey leaves (wear gloves because fine hairs on the leaves can be an irritant).

1.
Chopping up your nettles and weeds will increase the surface area, and accelerate the rotting process.

2.
Fill a bucket with water, or use your water butt.

3.
Add the weeds/comfrey to a bucket above of water – around 1kg of nettles to 10 litres of water, or 1kg comfrey to 15 litres – and place a lid or cover over the top.

4.
Leave for a couple of weeks – you will know they are rotting when they start to smell.

5.
Stir and replace the lid.

6.
Leave another week or so.

7.
Use by spooning off the liquid, or strain out the rotting matter, which should be brown and degraded.

8.
It can be stored in a labelled plastic bottle in a cool, dark place.

9.
To use, dilute the tea 1:3 with water.

MAKE A WORMERY

Wormeries vary in complexity from a simple, single box with a lid to expensive kits and contraptions.

Here is an easy homemade wormery, which you can simplify further by leaving out some steps and using just the lower box on its own.

1.
Take three stackable boxes (black is best, because worms are light-sensitive, but black boxes may be harder to find).

2.
Drill holes around the tops of two of the boxes and also across their bottoms (the third box is to contain the fluids that come from the worms, known as liquid exudate).

3.
Drill a large hole at the lowest point of the exudate box, large enough to take a plastic tap fitting.

4.
Add a layer of scrunched up newspaper to one of the aerated boxes.

5.
Then add a layer of fresh garden compost, and/or well-rotted manure.

6.
Then add a layer of kitchen scraps.

7.
Water it a little.

8.
Add worms.

9.
Cover with a lid or the second box.

10.
Feed your worms little and often with kitchen scraps and keep moist but not wet.

11.
Once the lower box is full, add some compost and newspaper to the upper box, and the worms will move up into it to be fed new scraps.

12.
Draw off the nutritious exudate ('worm tea').

13.
Dilute it 1:10 with water and use as a fertiliser.

14.
The worm tea can be stored in a labelled plastic bottle in a cool, dry place.

15.
You can then use your new compost on the garden or in your pots.

PROTECTING YOUR SOIL

The good news is that although soil is delicate and easily damaged, there are steps that we can easily take to protect it.

MULCHING

Mulching is the application of a protective barrier to the surface of your soil. It happens in nature when leaves fall and accumulate on the ground below a plant. This layer of leaves breaks the force of falling raindrops, reducing the risk of water erosion, and stops light soils being carried off by wind. A thick layer of mulch also traps moisture and warmth within the soil, improving growing conditions and plant health. The final benefit for gardeners is that by preventing light from reaching dormant weed seeds in the topsoil, mulch reduces the germination of weeds around your cultivated plants.

The natural mulch of leaves will eventually rot down and become incorporated into the soil itself. This is where some confusion may arise between the terms mulch, compost and fertiliser, because as leaf mulch rots, its organic matter becomes compost for the soil and in doing so, it also adds nutrients and so acts as a fertiliser. Isn't nature amazing – mulch, compost and fertiliser: all in a fallen leaf!

The protective soil-stabilising, moisture-retentive and weed-suppressant properties of a mulch can also be achieved by non-organic material, such as gravel. However, you won't then get the compost and fertiliser benefits of rotting organic matter.

So, mulching materials include gravel, ornamental glass chippings, crushed seashells, as well as ornamental bark, garden compost, straw and leaf litter. Farmers even use plastic sheeting as a mulch on crops (see Microplastics, page 34).

CHOP AND DROP

Collecting up your plant cuttings and leaves in autumn takes effort. You gather it all up, place it in a corner to rot down and then spread it all out again a few months later. Why not just leave it where it is? Chop and drop, as the name suggests, is simply the method of chopping down your plants and then leaving the cuttings there on the bed where they fall. One small caveat to this – if your plant is diseased or carrying pests, it is best to remove its cuttings and leaves to general waste.

WATERING TECHNIQUES

Soil in the natural landscape usually has a covering of organic matter to buffer the physical effects of a downpour. That protective shield begins high up with the broad umbrella of the tree canopy.

Every time a raindrop bounces from a leaf and falls to the next layer of foliage it loses energy, until it plops gently to the ground. Finally, on the ground a layer of either living plants, fallen leaves or rotting vegetation absorbs the impact of the droplets and the water is absorbed gradually into the soil.

This ideal scenario is what we need to aim for when we water our garden artificially. It is upsetting to think that our watering could be damaging the soil, and thus the plants, rather than supporting them.

There are a number of simple rules and techniques to bear in mind if you want to limit the damage of watering. First, I must say that it is always better to water as little as possible. Water is a precious and finite resource, not to be wasted.

Even if you have an abundance of recycled rainwater to use, applying it artificially is never quite the same as natural rain; it is difficult to get it right: the amount, the intensity, the timing. It is another reason to plant the types of plants that are right for your soil and situation; plants that, once established, can survive periods of drought. However, even the most well-planned planting scheme will need watering to establish the plants, and see them through the occasional extreme event.

Timing is the first thing to consider: do your plants need watering right now? There are times in a plant's life or annual cycle when it is particularly vulnerable, such as early in the growing season when new soft leaves are emerging, or when a seedling is germinating. At the end of the growing season when a plant is about to enter its yearly dormancy, it may also not need watering; perhaps it is taking on autumn colour or dropping its leaves a little early as a natural response to drought, and it will be fine come the spring. Plants have protective mechanisms, and each plant has its own resilience.

So, if one plant needs extra water, it does not mean that they all do. It will even vary depending on the position of a plant in your garden. A thirsty plant in the cool shade of a taller companion will survive drought better than the same plant positioned in full sun. So, it's really about getting to know your garden and the plants in it. How the soil, microclimates and natural characteristics of each plant work together and how you tailor your support to the individuals.

Once you have decided to water, choose the coolest time of day; otherwise, the water can just evaporate. Morning is best because fungal spores tend to be released in the evening and these can be spread by your watering efforts.

These are also good reasons to avoid sprinkling up into the air. Water down at the base of the plant, using a fine spray so the water can be easily absorbed rather than running off across the surface away from the plant.

Hand watering is best because you can assess each plant's individual need. Make a mental note of how much water you are giving it and how it responds. Time spent hand watering your plants is a great way to get to know them.

This is not something you need to do every day. It is better to give your plants a good drenching a couple of times a week rather than a little daily. This is because shallow watering will only wet the very top

of the soil and so the roots of your plants will grow toward, and stay, on the surface level of the soil. We want to encourage the roots to dive deep down to the cool, refreshing water of a generous application. A good way to get to know how much is enough is to press your finger down into the soil, or dig down a little, to see how far the water has penetrated. I guarantee you will be surprised the first time you do this. After a few tries, you will get to know how much watering your soil needs.

This will be affected by the thickness of your mulch layer. You apply a mulch to keep moisture in the soil, but it may also keep some rain out. For this reason, it is a good idea to water well before applying a thick layer of mulch. If the mulch is thick, it is usually the upper layer that will dry and lower layers will stay moist. However, prolonged drought may dry all your bark, for example, and then you will need to consciously rehydrate the mulch as well as the soil beneath it – giving it all a very thorough watering

Of course, if you have a large garden, it may not be possible to hand water, so you will need an automated system. There are a great many to choose from, but the simplest to install, most economic and least problematic to your soil is some kind of porous or leaky pipe system.

As the name suggests, these systems have pipes that ooze water from them. If they are laid across the surface of your soil and around the roots of your plants, they will gently release moisture right where it is needed, and with a timer exactly when necessary.

Experience has taught me that you do still need to be vigilant. Some plants

Top: The force of a hose can damage the surface of the soil. Water gently and deeply at the base of your plants and only when they actually need it.

Bottom: Check if your plants need watering by feeling below the soil surface with your finger.

may need more or less than the pipe is delivering, so be ready to adjust the system and supplement as required. In this way you and your plants get the best of both worlds; you save time and your plants get water when they need it as well as a little of your individual attention.

Below: The cheapest DIY irrigation is easily fitted porous pipes, on a timer.

PAVING AND RAINWATER RUN-OFF

In Britain, we have a great deal to thank the Victorians for. At a time of prosperity and innovation, they built an infrastructure of drainage channels that revolutionised the nation's health. Their system has been carrying sewage and rainwater away from our homes for over a hundred years, but even the thoughtful Victorians did not anticipate the extent to which our towns and cities would grow.

Why is it that when I walk into my village on a stormy day, at a certain spot I know I will see a gulley popping up out of the road surface, atop a gushing fountain of dirty rainwater? Why are the drains failing?

Nature is well equipped to deal with rainwater. When rainwater falls on to soil or plants, it seeps through the layers of the ground and eventually back into aquifers or rivers. Where we have laid man-made, hard, impenetrable surfaces, gullies and drains are needed to carry the water away.

The problem is that now, in many parts of the United Kingdom, the areas of manmade concrete, asphalt and stone are too great for our old drainage system to cope with. If you live in a new, well-designed suburban area with adequate drainage facilities, you might be thinking, 'I'm alright, Jack!' Surely, the answer is to build bigger drains!

However, we really need as much as possible of the rain that falls to soak into the soil, so it can return to the delicate and complex local natural cycle and not be diverted away either for resource-intensive treatment or to be sent into waterways.

Most modern paving surfaces are impermeable not only to water but also to air. This means that the organisms in the soil beneath cannot breathe and will die. The soil beneath concrete, asphalt and probably your patio is dead.

Some natural paving materials, such as hoggin and gravel, are porous to water and air. Traditionally, these would often have been laid easily and cheaply, and so if they wore thin, they would simply be topped up. In our search for permanent solutions

that are suitable for heavy modern traffic, we have used more and more engineered paving systems designed to resist the weathering elements.

If you have a typical paved patio or driveway, it was probably laid by first compacting the native soil to squeeze out all the air, then laying and compacting a thick layer of stones, possibly followed by a layer of concrete, and then finally the decorative slabs that you see. It is completely impervious to air and water, so the soil dies and the rainwater needs to be managed: drained away.

In 2010 the UK government passed a law restricting the use of impermeable

Above: Loose aggregates allow rainwater to filter back down into the soil.

Right: Gaps in paving improve rainwater circulation and can aerate the soil, supporting soil life beneath.

paving in front gardens, and for some time newbuilds have been required to process their rainwater on site rather than adding to the burden on public systems. Along with other governments worldwide, it is trying to address these issues by encouraging and mandating Sustainable Urban Drainage Systems (SUDS).

Above: A sample of a new permeable concrete that allows water to filter through.

The ideal is to keep hard surfaces in your garden to the bare minimum and 'green' them wherever you can. Where you do need to have impermeable paving, try to place some planting or lawn nearby, so that you can shed the rainwater on to the 'soft' areas rather than into drains. Even a narrow green drainage strip will help.

Many people will feel that, for practical reasons, they need areas of hard, dry paving in their gardens for people or cars. However, if you would like to consider

sustainable solutions, they are becoming increasingly available and popular.

You could revert to those traditional natural materials and methods of laying and accept the occasional maintenance requirements as part of the deal. Or you could choose one of the modern, engineered solutions.

It is important to remember that a paved area has a foundation, so in order to maintain its permeability, the base must be as permeable as the surface. This is often forgotten. Research is now showing that an advantage of this is that many of the permeable aggregate layers are also filtering impurities out of rainwater run-off: a quality that might be harnessed, particularly in polluted urban areas and roads.

Paving and driveway design can get complicated and is beyond the scope of this book; however, the take-home message is the more green and planted you can make your outside space, the better.

WHEN NOT TO DIG

As I write, in my area we are having a record warm and wet winter. It is tempting to grasp the opportunity along with my garden fork and get ahead of the weeds, which don't seem to have stopped growing this winter. However, I resist.

When I am planting a newly designed garden, human schedules often dictate that we clear beds, dig and plant as fast as we can, rather than using best judgement. In your own garden, you will have more scope to do what is best for your soil.

Working the soil in the cold or winter can have long-lasting detrimental effects on its structure, particularly if your soil is clay, with its delicate structure and tendency to hold on to water and cold. Once the soil structure is damaged, it takes a good while to recover.

It may seem warm to me now in Sussex in February, but soil temperature lags behind air temperature; it is still very much winter down there. It is better to wait until the warmth of spring to dig sandy soil. For clay soil, in fact, the best time is autumn, when you can leave the clods to be broken up over winter, by the frost. However, it is okay to dig clay soil in spring, as long as it is not cold and wet. The trick with clay is to find that sweet spot between cold, wet soil and the moment it becomes hard as a rock in summer.

Below: When soil is frozen, digging can damage its structure. Best to put your feet up, have a warm drink and wait for spring.

NO-DIG GARDENING

A resurgence of an old gardening method that is making 'lazy' gardeners (like me) very happy.

THE PRINCIPLES OF NO-DIG GARDENING

You may have heard Aristotle's saying, 'Nature abhors a vacuum.' Every gardener will tell you that is certainly true of soil. As soon as you have cleared your vegetable bed of weeds and sit back to admire your work, the weed seedlings will appear as bright green shoots through the fine tilth of bare soil.

This may be frustrating for the tidy gardener, but as with all things in nature, it is part of the complexity of systems that maintain environmental harmony.

There is a whole seed bank stored in the upper layers of your garden soil. Billions of mostly weed seeds waiting for the right mix of moisture, warmth, air and light to spring into action. Every time you dig the soil over, you are bringing hundreds of new seeds up on to the surface where they rapidly germinate and we wonder why a gardener's work is never done!

Infuriating as this might be, it is not the most problematic part of your digging. When you dig, you disrupt the structure of your soil. Mini beasts and microorganisms have habitats within the soil, often preferring particular layers within its structure; when you dig, you literally turn their world upside down! Some create burrows and pockets which help to aerate the soil and improve drainage. Plant roots reach well beyond the spread of their leaf canopies and the hyphae of fungi reach far through the soil, all macroscopic and microscopic networks essential to soil health. The less we dig, the better.

If, instead of digging, we simply lay organic plant matter on to the soil surface, the weathering processes and 'rotters' of nature will degrade it, and an army of mini beasts will draw the material down into the soil.

CONTAINER GARDENING AND RAISED BEDS

With the layering of the no-dig method, over time the soil level in your beds will rise a little, so you will end up with flat-topped mounds of lovely, healthy, crumbly soil.

As a designer, I am drawn to clean, intentional lines and I struggle with too much 'shagginess'. For these reasons, I prefer to define my planting beds with wood edging, but it is not really necessary; it's an aesthetic, personal thing.

It is wise, however, to define paths between your beds, because they will help you to avoid walking on and compacting the soil. The paths can be as permanent or temporary as you like. Often with vegetable beds, people prefer to switch things up by moving the beds and changing their size and orientation each year, so straw or bark chippings is a good, porous solution..

Edging does have the advantage of allowing you to raise the height of your beds, which not only reduces bending and potential back pain, but can also look very smart. One disadvantage, especially in damp climates like the United Kingdom, is that slugs and snails love sheltering in the nooks and crannies created by edging, and woodlice love it if the timber starts to rot.

Thinking again about mimicking the natural structure of soil, underneath the topsoil, where most of the plants grow, there is usually a deep subsoil layer. So, there is little point having a very tall raised bed filled with topsoil. In fact, a topsoil layer deeper than about 30cm/12in in a raised bed or container will tend to compact under its own weight. It is better to have a drainage layer at the bottom of your raised bed, or a layer of unrotted organic matter, such as garden waste, wood chips, or even kitchen vegetable scraps.

Below: The 'no-dig' technique involves avoiding soil disturbance by layering organic matter on top of undisturbed soil and keeping to footpaths between the beds.

BUILDING A RAISED BED

———

A cost-effective and simple raised bed is made with timber but you can use any material. Thick timber sleepers have the advantage that they are chunky enough to contain the soil and keep their shape. I use cardboard to suppress the weeds, wire mesh to keep rodents out, and either cardboard or woven geotextile to stop the soil from oozing between the gaps.

1.
Choose your spot and mark out the area and shape of your new raised bed.

2.
If it is currently grass, mow it, or you can build your raised bed on to poor soil, concrete or gravel.

3.
If you have perennial weeds in the area, you can keep them at bay by placing a permeable layer at the bottom of your bed. Cardboard, hessian or newspaper will eventually rot, whilst woven geotextile will act as a permanent barrier.

4.
Build the sides of your raised bed.

5.
A fine steel mesh across the base of your bed will help to keep out burrowing creatures.

6.
You may want to line the inside with a barrier material to prevent soil leaching out through the gaps in the wooden sides (see choices of material above).

7.
If your raised bed is taller than 30cm/ 12in, in the bottom place a mix of woody material of different sizes for drainage (this is also a good idea if your bed is on a hard surface like concrete, which will not drain freely).

8.
You can fill the bed either with a good compost and topsoil mix, or with different layers that will rot such as vegetable scraps or woody material.

KEYHOLE BEDS

A few years ago, I went to Tanzania to help establish a kitchen garden and, as is so often the case, I learned as much information as I shared. I had not seen a keyhole garden before, but I soon recognised the principles of no-dig.

The bed is shaped like a keyhole; in fact, my hosts called it a 'love garden' because, with a little imagination, one might see a love heart in its shape. It is cute but also practical because it allows access to the centre of its circular shape without having to walk on the soil. It can be constructed from any material. This being Africa, we used what was to hand: some clay bricks leftover from a recent construction. At the centre of the bed, we placed an old wicker basket and filled it with kitchen scraps, some manure and woody cuttings. Then around this, filling the rest of the space, we layered woody cuttings, compost, manure, green garden clippings and weeds. The idea is that you water into the central well, which rots the organic matter at the centre of the bed and a nutrient-rich leachate oozes out into the rest of the bed, which in its turn rots to create a rich growing medium. Genius!

CONTAINERS

A planting container is basically a small, ornamental raised bed. The principles are the same; however, there is a wider choice of styles and materials.

Gardeners often used to talk about 'crocks' for drainage. This was when the general use of clay pots in gardens resulted in a common abundance of broken bits of pots that could be used at the bottom of any growing container for drainage. Now we struggle for crocks, but people often use 20–25mm/¾–1in diameter pea shingle or recycled scraps of polystyrene; anything really that will allow drainage and prevent the fine soil or compost blocking the drainage holes.

However, a controversy has since arisen. Research has now suggested that gravel (or pea shingle) at the bottom of your container may, in fact, hold on to water and reduce drainage. It is thought that by capillary action, water is held around and between the stone particles, thus creating the very opposite of the desired effect.

That being said, it is still important to try to prevent the drainage holes from becoming blocked by soil. Carefully place a mix of large and small brick pieces, tiles and stones around the hole and then continue filling. With a container you can use the same fill options as with raised beds, but do ensure that materials are finely chopped so they decompose quickly.

Right: This example of a keyhole bed or 'love garden' has a shape that enables the gardener to get close to the planting without walking on the bed. Some designs incorporate in-ground composters or irrigation points to allow nutrients and water to seep out into the soil.

HÜGELKULTUR

Hügelkultur is German for hill, or mound, cultivation.

A woody base allows air to circulate through the mound and also the free drainage of water. The logs and branches also offer a habitat for a wide variety of animals.

In the first year, stick to planting shallow-rooted plants, so they can establish in the top layer of soil above the woody base.

The woody material holds moisture at the base of the mound, but it may dry out at the top, so you may need to water occasionally.

As the wood rots, it releases heat for the growth of your plants and also releases nutrients. Unfortunately, this rotting of carbon-rich wood may also eat up nitrogen, so you may need to supplement your plants with a nitrogen-rich organic fertiliser.

A disadvantage is that as the material rots down, the height of the *hügel* will drop, but this is easily compensated for by adding layers of mulch and compost to the top.

The best time to start your *Hügelkultur* is in the autumn, to allow it time to develop before planting in the spring.

Right: A *hügel* bed is not a very pretty garden feature, but it harnesses the natural processes of rotting organic matter to boost the growth of your vegetables. It provides habitat for insects and animals, heat generated from the decomposition, moisture retention of organic matter, and eventually the release of nutrients from old plants to new.

1.
Choose a spot and mow any existing grass.

2.
Remove the turf and topsoil and store for later reuse.

3.
Dig a depression in the ground and line it with cardboard, newspaper or hessian to contain the growth of any invasive perennial weeds.

4.
Place some large logs on the barrier. This is the base of your mound; it will rot very slowly, creating heat for your growing plants and eventually become rotted organic matter.

5.
Place smaller twigs over and between the logs, then compact it all down and water.

6.
Place more small twigs and leaves in and around the woody mound, compact and water.

7.
Repeat this layering, watering each time.

8.
Once the mound is as tall as you want it, invert your turves to cover the woody material and cover that with compost, manure and topsoil.

9.
Generally, a layer of about 30cm/12in will block light from the turves and prevent the grass regrowing. If it is difficult to achieve this, or if you suspect weeds may grow on your mound from the soil that you have used, cover and leave over the winter months. The rotting process will start and the weeds will die, so that in the spring you will have a good, warm growing area ready to start gardening.

RAIN GARDENS

All styles of rain garden use the principle of water attenuation: holding back an influx of rainwater to allow it to gradually seep out.

PRINCIPLE OF RAIN GARDENS

There is a finite amount of water on our planet. It constantly cycles from land and seas into the atmosphere through evaporation and back down again as rain. When water falls on to the ground, it begins its long journey, through the layers of soil and rock towards water courses, which ultimately return it into the atmosphere. In any one location, the routes that rainwater will take are determined by the physical and chemical composition of the land, as well as its topography: hills, valleys and undulations. In some places, rainwater moves quickly through the landscape and in others it

collects, sometimes as temporary flooding of previously dry land, and sometimes as permanent ponds, lakes and rivers with varying water levels. There is a delicate balance between dry land and wetland, which is constantly in flux, always adjusting to the myriad complex requirements of the ecosystems.

Fundamental to this is the process of attenuation. This is the capacity of soil to absorb water and hold on to it for a time. The rainwater is held in the spaces between soil particles, and then gradually released, as the land drains, by the action of gravity, while some water is absorbed into the atmosphere. This swelling of the soil with rainwater is a protective mechanism which reduces the risk of local flooding and gives time for the landscape and its inhabitants to adjust as the water is gradually released.

A second related protection is the collection, locally, of excess rainwater in ponds, lakes and rivers. Natural bodies of water have a capacity to contain excess water, to a point, before bursting their banks. The shape of land, with its dips and hollows, has evolved to collect rainwater, thus allowing it time to seep away into the soil.

A rain garden is an artificial facility that you can easily create in your own garden to collect water temporarily, and thus give it time to seep away into the soil gradually and naturally.

It is a way of directing excess rainwater to a place in your garden where it will not cause problems and hopefully can be used to good effect in an aesthetically pleasing design.

MOIST SAND WITH PORES

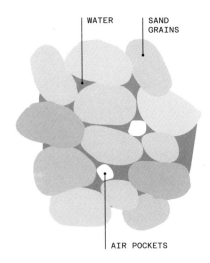

WATER

SAND GRAINS

AIR POCKETS

Opposite: This decorative raised bed is taking rainwater from the gutters of the house instead of allowing the precious resource to flow away down the drain. The rainwater is also flowing into an ornamental pond beneath the grill set into the paving.

Above: Soil particles have air pockets between them and these can become filled with rainwater. In the short-term, this does no harm because gradually, under the pressure of gravity, the water seeps out. The size and number of the air pockets will vary. If there is too much water for the soil to accommodate, then waterlogging or flooding may occur.

DIFFERENT STYLES
OF RAIN GARDEN

A concrete storm drain, designed to contain the waters of a flash flood, carries the water away from harm, but it does not harness the protective, absorptive qualities of the soil, because it is lined with impermeable concrete. A rain garden differs in that it not only contains the rain, but also allows it to gradually dispel.

Another advantage of a rain garden is that, because it can be planted with a variety of water-adapted plants, it increases the biodiversity of your garden and can create a different habitat for wildlife.

At its very simplest, a rain garden can be just a dip in the lawn where you know the water from a heavy downpour will collect and gradually drain away.

If you are feeling more adventurous, you could plant that dip with plants that will not mind getting flooded for a short time and also will not mind drying out in summer. Luckily nature already has such a group of plants: those that happily grow along the edges of ponds and streams – 'marginal' plants.

This planted type of rain garden looks pretty much like any other flower bed – just a bit wetter sometimes.

It is also possible for your rain garden to be more of a water feature, if you design it so that the water rises in the area to the extent that is visible for a while. Then it will look like a pond or stream when it is raining and a 'dry-stream' of decorative rocks and pebbles in summer. Again, marginal planting and absorption through the soil are essential to the function of this planting scheme.

The mechanisms for getting water into and out of your rain garden can vary in complexity. The simplest is for gravity to carry the rainwater down a slope into the depression and then to seep out into the soil. This is known as a self-contained rain garden.

Another easy way to deal with the rainwater flowing from the roof of your house is to divert a downpipe away from a gully in the ground and off towards your rain garden.

It is important to ensure that your soil is permeable enough to allow water to drain. This can be easily checked with a simple test.

Dig a hole approximately one spade deep and fill it with water. Allow that water to drain out, thus saturating the soil in the hole. Once the water has gone, fill it again and time how long it takes to drain away. If it takes less than six hours to empty, you have the right conditions for a rain garden. If not, you can always improve the drainage by adding grit to the bed, although you will need good amounts to achieve approximately a 30:70 to 50:50 mix of grit to soil.

If you anticipate a lot of water in your rain garden and you would like to help it to flow away more quickly than natural seepage will allow, it is possible to enhance the rate of drainage by adding a perforated drainpipe. This could drain away to another rain garden, a traditional underground soakaway, or even back into the mains drainage system. This is an under-drained rain garden.

Not everyone has the space for a rain garden bed, so a compact and easy alternative way of diverting rain from the mains system and increasing biodiversity in your garden is to install a rain garden container next to the downpipe. This is a pot or tub filled with free-draining soil and planted. At the bottom of the container, you should have a layer of gravel for drainage. It is also important to have an overflow pipe, which protrudes from the surface of the soil to capture excess water and carry it out of the bottom, away to a drain.

Opposite: The simplest form of rain garden: an outflow pipe into a vegetated ditch.

Above: A naturalistic rain garden designed to look attractive even when the water has dried out.

CREATING A RAIN GARDEN PLANTER

There are as many rain garden planters as there are containers to adapt. Your planter could be quite formal and ornamental, or it could be rustic and 'upcycled' like the one I created on my allotment (opposite).

Because each rain planter needs to have an overflow, you could rig up another next to the first to take that overflow. However, do bear in kind that your second planter will need to either sit at a lower level or have the water just flowing into its base for the plants to soak up from beneath.

Choose a generous planter that will fit neatly near an accessible downpipe. Shorten your existing downpipe and duct-tape it over the top of your planter.

1.
Install a drainpipe so its top will be just above the level of the soil.

2.
Drill a hole in the bottom of the planter and ensure the excess water will drain from the bottom of the planter to a suitable gulley or drainage point.

3.
Fill the bottom quarter of the container with 20–25mm/¾–1in diameter pea shingle or other drainage material.

4.
Cover the drainage layer with a permeable separating layer – this could be reused garden sacks pricked with holes, woven geotextile or a biodegradable material which will, in time, rot away.

5.
Fill the planter with soil and plant.

PLANTS FOR RAIN GARDENS

The most important plants in your rain garden are the 'riparian' or 'marginal' plants that are adapted to the alternating wet/dry/wet conditions of a stream or pondside. However, in order to create a pleasing feature which looks settled in your garden, you may want to blend the edges of your rain garden up on to dry land, where xeric (dry soil adapted) plants will thrive.

Here are some of my go-to, dependable plants that will tolerate a degree of flooding. It is good to plant a mixture of taller types with lower, creeping plants that will run between the others, covering the soil.

PLANT	DESCRIPTION	NOTES
 **Bugle** *Ajuga reptans* ◐	A low-growing creeping plant with semi-evergreen leaves and small spikes of purple-blue flowers in summer.	UK native It is fairly resilient for blending on to drier land, but much happier if it doesn't dry out.
 **Astilbe** *Astilbe* 'Bressingham Beauty' (× *arendsii*) ◐	60–90cm/24–36in tall perennial with cut leaves and fluffy plumes of pink flowers in July, August, September.	The *arendsii* types like moist soil, whereas the *chinensis* types will tolerate drier conditions, so can be used to blend your scheme on to dry land.
Bistort *Bistorta amplexicaulis* 'Fat Domino' ◐	A well-behaved perennial, 90cm/36in tall, with big, fat, poker-shaped red flowers in July, August, September, and October.	The leaves fall into a heap which last the winter on the ground, thus providing a habitat for wildlife during cold months.

Palm sedge

Carex muskingumensis

☉

60cm/24in tall mounds of bright evergreen, grassy-like leaves.

In June and July has pompom-like straw-coloured flowers.

Great for maintaining winter structure in your rain garden.

The leaves will last through the winter, when you can comb out the tatty leaves and dead flower-heads.

Tufted hairgrass

Deschampsia cespitosa

⦂⦂

50cm/20in tall evergreen hummocks of fine grass with airy flowers in summer.

UK native

Joe pye weed

Eupatorium maculatum 'Riesenschirm'

(syn. *Eutrochium* 'Riesenschirm')

⦂⦂

2m/7ft tall perennial form of eupatorium, with dark purple stems and clumps of wine red flowers in August, September, October.

Loved by pollinating insects.

Yellow iris

Iris pseudacorus

☉

70cm/27in tall straps of pointed leaves with bright yellow flowers in summer.

UK native

☉ Sun ● Shade ◑ Sun/partial shade   Any

PART FOUR :
WHAT TO PLANT FOR A HEALTHY SOIL

HOW PLANTS HELP THE SOIL

Plant growth is dependent on a healthy soil and, as with all natural ecosystems, the plants have a great deal to offer soils in return.

SOIL DECOMPACTION AND AERATION

Small-particled clay soils are particularly prone to damage, compaction and loss of air. All plants help to prevent soil compaction by improving the levels of organic matter in the soil, and thus the soil structure. Increased organic matter and roots within a soil help the particles to clump together into larger aggregates that have more air between them. This helps to make the soil crumblier and lighter, and to hold a good amount of water.

Natural improvement of soil structure takes time. Fortunately, there are plants with roots that can penetrate heavy clay and rapidly break it up. Some have thick, deep 'tap' roots, while others have a dense network of determined fibrous roots.

If you are planting to improve your soil structure, it will usually be temporary, so beware of perennials with robust root systems and prolific habits (such as dandelion) that may grow out of control.

Perennial clay-busters and deep tap roots will need to be dug up. With heavy, clay soil it is unlikely that you will be able to pull them out without disturbing the soil. So, if you want to use a no-dig strategy, I suggest you plant annuals to improve soil structure.

Be careful when your annuals are due to flower; it is generally better to cut them down before they set seed and return the next year. An exception to this is tender annuals, such as white mustard, which will hopefully be killed by winter frosts, saving you the trouble of removing them.

WELL-BEHAVED AND ANNUAL 'CLAY-BUSTERS'

Tap-rooted:
- Potato (*Solanum tuberosum*)
- Daikon radish (*Raphanus sativus*)
- 'Bocking 14' comfrey (*Symphytum × uplandicum*)
- Turnip (*Brassica rapa*)
- Beetroot (*Beta vulgaris*)
- Artichoke (*Cynara scolymus*)
- Lupins (*Lupinus* types)

A WORD ABOUT COMFREY
Comfrey has deep roots and most varieties self-seed abundantly. 'Bocking 14', being a sterile cultivar, is less problematic, but it is wise to divide it every few years to prevent its roots becoming too established.

Fibrous-rooted:

- White mustard (*Sinapsis alba*) – half-hardy annual
- Hungarian grazing rye/forage rye grass (*Secale cereale*) – hardy annual
- Winter field bean (*Vicia faba*) – hardy annual
- Yarrow (*Achillea* types) – perennial
- Red clover (*Trifolium pratense*) – perennial

A WORD ABOUT RYE GRASS:
Secale cereale goes by different common names. It is one of the best cover crops for adding organic matter to your soil and breaking up clay soil. Like winter field bean, it also has the advantage that, if sown in autumn, it is hardy enough to overwinter. The leaf cover remains during the cold, wet season, protecting clay soil from compaction and erosion.

NITROGEN FIXATION (LEGUMES)

Nitrogen atoms are building blocks of proteins, which are essential to plants and animals for making their cells and biological functions. Nitrogen occurs in nature in various forms, either as nitrogen gas in the air (N_2), or as compounds such as ammonia, ammonium, nitrates or nitrites. These compounds are more or less soluble (dissolvable in water), which affects how easy they are for organisms to take up and use. All the nitrogen that we humans need to live has to come to us from plants, but how do plants get their nitrogen?

Plants and animals need to turn nitrogen gas into a soluble form in order to use it.

Plants are dependent on bacteria to 'fix' gaseous nitrogen for them and these 'nitrogen-fixing' bacteria live in the soil. The bacteria have an enzyme, 'nitrogenase', which enables them to combine gaseous N_2 with hydrogen (H_2,) to make ammonia (NH_3)/ammonium (NH_4+) and nitrates (NO_3-) via nitrites (NO_2-): all compounds that plants can absorb.

Nitrogen-fixing bacteria can move freely in the soil, releasing nitrogen generally out into the soil, or they can connect with the roots of plants, either loosely or in a specific, close relationship.

Plants grow tiny hairs on their roots, which stretch out into the soil between soil particles. These root hairs absorb soluble inorganic ammonium. The root hairs also attract nitrogen-fixing bacteria.

If loosely associated with the plant, the bacteria release ammonia/ammonium and nitrates into the soil near the root hairs, which absorb the essential compounds. The plants then use these inorganic forms of nitrogen to make chlorophyll and other forms of organic nitrogen, which are essential to photosynthesis and metabolism.

Some plants have found a way to 'farm' nitrogen-fixing bacteria by growing nodules on their roots that attract bacteria with their protective environment and carbohydrates produced by the plant. The bacteria form a symbiotic relationship with the plant. This enables such plants to survive in nitrogen-poor soils where others would not. They have a bolt-on nitrogen factory.

When nitrogen-fixing plants die, the high concentrations of ammonia/

ammonium and nitrates in their root nodules are released into the soil as they decompose, thus raising the nitrogen content and fertility of the soil. So, when you grow nitrogen-fixing legumes, the benefits to the soil and plants are only gained when that plant dies. It is, of course, essential to leave the roots in place as well to rot down and release the nitrogen from their nodules.

Studies have shown that alfalfa and red clover are the best nitrogen fixers,

but the choice depends on many factors: other requirements and also horticultural conditions suited to each plant.

SOIL DECONTAMINATION AND WATER FILTRATION

'Phytoremediation' is a fascinating and recently developing area of science that uses plants to clean up soil and water. Some plants (and their associated microorganisms) take up harmful contaminants and either store them or transform them into less toxic compounds.

Probably the best-known example is using the roots of reeds (rhizofiltration) in reed beds to filter pollutants and phosphates out of 'grey' and 'black' water. This can be done on a small scale in your back garden to manage just your own waste water or on a massive communal scale with specially constructed wetlands that also offer recreational value.

Phytoremediation is the general umbrella term. However, other terms can be used to describe the particular process: phytoextraction, phytoaccumulation, phytostabilisation and phytodegradation.

Poplar and willow trees, sunflowers, barley and sugar beet can all be used to take up salt to reclaim land from the sea. Hemp and mustard take up and accumulate heavy metals. Genetically modified *Panicum virgatum* or switchgrass, along with the bacteria that it holds around its roots, is being used to clean up organic explosive contaminants on military testing sites.

The plants have limits to what they can do, but they seem to be a relatively cost-effective and important tool in our fight to clean the planet of our waste materials.

Above: Round nodules on a leguminous (pea family) plant.

Opposite: *Panicum virgatum* (switchgrass).

SOIL STABILISATION (GROUNDCOVER PLANTS)

Bare soil is vulnerable to erosion by wind and rain. The soil surface is subject to evaporation of its moisture, making it even more vulnerable, as when dry its smaller, lighter particles are easily lifted into the air. On a large scale, this constitutes soil erosion.

Above: Bare soil lacks fibrous roots that knit its layers together, making it vulnerable to water and wind erosion.

Opposite: Many grasses are pioneer plants, colonising bare ground. They have fast-growing roots that rapidly infiltrate a soil, especially a loose soil, and stabilise it.

Water, too, can erode a bare soil. Whether it is the flow of streams and rivers, or the gravitational pull of rainwater flowing through and across the soil, its particles will be carried away, along with the nutrients therein.

The sooner bare soil is replanted the better. The fibrous roots of plants bind the soil particles and layer them together, stabilising them against the pull of wind and water. The leaves of the plants act as an umbrella: a weather shield protecting against the fall of raindrops and the forces of the wind.

Some plants offer a more effective protection than others. This will be due to their speed of growth, how fibrous their roots are and how thick their leaf cover is.

Grasses are fast and efficient soil stabilisers, as demonstrated by any child who has rolled down a grassy bank in dry weather. Lawn turf grows roots within a few weeks and is the quickest and easiest way to stabilise a hillside or slope. Spreading ornamental grasses will take a little longer, but once established will knit even a light, sandy soil together well.

Robust, low-spreading shrubs, such as *Cotoneaster dammeri*, adjust their habit to the terrain and quickly form a protective leaf canopy over the soil.

Trees, particularly fast-growing pioneer trees, are large-scale soil stabilisers. Whether it is silver birches on sandy heaths or willows beside a river, their wide-spreading root systems can hold together large areas of vulnerable topsoil.

TOP SOIL-SAVING PLANTS

Here are my personal top plants, which have more than one useful quality for saving your garden soil.

PLANT NAME	SOIL-SAVING PROPERTIES
Alnus glutinosa	Water retention and nitrogen fixer
'Bocking 14' comfrey	Decompaction and compost tea
Potato	Clay-buster and edible
Red clover	Cover crop and nitrogen fixer, bees love the flowers
Cotoneaster dammeri	Soil stabiliser, will grow anywhere and bees love the flowers
Bergenia types	Groundcover, weed suppressant, soil-stabilising ornamental
Taxodium distichum AGM	Large tree for future warmer climate change Wet and dry soil tolerant
Ligustrum japonicum (and other species)	Mitigates rain and eases flooding
Wildflower lawn	Loads of biodiversity in a small space, practical in the garden, protects the soil

PLANTS FOR DIFFERENT SOIL TYPES

LOAMY

PLANT	DESCRIPTION	NOTES
Agastache 'Blackadder' ⊙	Deep purple 'bottle-brush' spires July–October. Loved by pollinating insects. Drought-tolerant, aromatic leaves, good for a mediterranean garden.	*Agastache* are short-lived perennials, which may disappear after a cold-wet winter. They vary in their frost-hardiness, with *A. rugosa*, Agastache rupestris, 'Blackadder' and 'Blue Fortune' AGM among the hardiest.
Convallaria majalis ⦂	Highly fragrant, delicate white bells 15cm/6in tall in April/May. Spreads rapidly and indefinitely if happy.	Can be tricky to establish. Loves moist, rich soil and cool root run. Best to allow it its own space; preferably in a raised position where the wonderful, but short-lived scent can be enjoyed.
Daphne odora 'Aureomarginata' AGM ◖	Evergreen, variegated, rounded shrub 1.2m/4ft tall. Small, pale pink, highly scented flowers January–April.	*Daphne odora* is temperamental, but worth trying to grow for its wonderful scent. It likes a sunny position, with rich, organic, moisture-retentive soil and it hates wind. If happy it will establish quickly, forming a rounded shrub with healthy, shiny leaves. It dislikes root disturbance.

 Sun   Shade   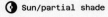 Sun/partial shade   ⦂ Any

Delphinium elatum 'Can-Can' AGM 1.6m/5¼ft tall ☉	Classic, showy, spires of purple frilly flowers with blue edges from mid-June.	Require staking and are prone to slugs. Protect emerging shoots as soon as the soil warms up in mid-spring.
Echinacea purpurea 'Rubinstern' ☉	Bright pink daisy flowers around central cone that persists into winter. Long-flowering June– September. and loved by pollinators.	*Echinacea* are notoriously tricksy due to their variable origins. Reliable, tested varieties include 'Green Jewel', 'Magnus', 'White Swan' and *E. pallida*. The new, double flowers and wonderful warm colours hailing from the USA may not overwinter the wet/cold in the UK. Prone to slugs and rabbits.
Fragaria (strawberry) 'Rosie' ☉	Good, reliable and tasty early cropper.	A rich, loamy moisture-retentive soil is essential for good flavour and juicy fruits.
Heuchera 'Paris' AGM ⠒⠒	Long-flowering, March–October, spires of fluorescent pink bells. Ornamental semi-evergreen leaves of green, marbled with white.	Amazingly varied group, well worth exploring for different leaf colours, flowers and cultivation preferences. Beloved by bees.
Imperata cylindrica 'Rubra' 40 × 30cm/16 × 12in ☉	Bright red upright blades of grass, which fade to lime-green at the base. The red of the leaves increases in intensity towards the end of summer, becoming spectacularly translucent.	Slow to establish; it takes a while to form what will eventually become a spectacular clump. Displayed well in a pot, but it does not like to dry out, so keep moist.

Paeonia 'Lafayette Escadrille' (Intersectional 'Itoh') ☉	Rich red, many flowered variety. Slow-growing, but worth the wait. 'Intersectional' cultivars have been bred to flower longer and not require staking.	They do not like to be disturbed. Avoid mulching or burying the developing shoots, which might rot; they flower better if they are 'frosted'.
Rosa 'Gertrude Jekyll' ◖	Highly fragrant, mid-pink shrub rose.	Roses have deep roots and so can withstand drought; however, they appreciate being watered in dry weather and will flower better if they are. They love a good dressing with manure and a deep soil rich in organic matter.
Solanum lycopersicum (tomato) 'Sungold' 2m/7ft tall ☉	Sweet, juicy, golden-orange small fruit.	One of the sweetest cherry tomatoes that can be grown outside. It thrives in a good, moisture-retentive soil. which can buffer the plant from drying out.
Zea mays (sweetcorn) 'Lark' 2m/7ft tall ☉	Sweet and less chewy than many varieties.	Mostly water, sweetcorn plants can lose a great deal of water from their leaf mass. They need a rich, moisture-retentive soil and benefit from extra mulch and mounding up of soil around the base to reduce root-rock in wind.

 Sun   Shade   Sun/partial shade   Any

CLAY

PLANT	DESCRIPTION	NOTES
Bistorta affinis 'Darjeeling Red' AGM ◑	Semi-evergreen, mat-spreading, low perennial with spikes of varying pink and red poker-shaped flowers. Leaves persist in most areas and are often flushed red.	Previously known as persicaria, this is a reliable groundcover plant.
Chaenomeles × superba (Japanese flowering quince) 'Pink Lady' AGM 2 × 1.5m/7 × 5ft ◑	Clusters of soft pink, open flowers along bare stems March–May. Aromatic and edible quinces in autumn.	A reliable and robust plant best trained against a wall, tolerates shade.
Epimedium 'Spine Tingler' AGM 30 × 30cm/12 ×12in ◑	Great evergreen groundcover. Prefers humus-rich moist soil, but once established is less fussy. Small, yellow flowers nod on wiry stems April–June.	There are so many wonderful epimediums to choose from. They should be better known. This one is evergreen, but also changes leaf colour from bronze, as young leaves emerge in spring, to a later green.
Escallonia laevis Pink Elle ('Lades'[PBR]) 1 × 1m/39 × 39in ☉	Small, evergreen shrub with mid-pink clusters of flowers June–July and then repeating through summer.	Previously robust, has succumbed to fungal leaf spot more recently, but I include in this list because it is so useful as a small hedge or clipped as flowering 'topiary' form.

Lonicera japonica 'Hall's Prolific' AGM 8 × 1.5m/26 × 5ft	Vigorous evergreen honeysuckle with highly scented cream/yellow flowers in summer.	Prune after flowering to keep under control and encourage flowering. Considered an invasive non-native species in Northern Ireland, it can be sold for garden use, but is illegal to plant in the wild.
Lonicera × *purpusii* 'Winter Beauty' AGM 2 × 1.5m/7 × 5ft	Straggly deciduous shrub. Small, cream, fragrant flowers in late winter/early spring.	One of my favourites for a lovely scented surprise on a winter's day. It doesn't look much, though, so tuck it away at the back of the border; it will still throw its lovely honeysuckle fragrance to the path, seat or doorway.
Magnolia 'Genie'*PBR* AGM 3 × 2m/10 × 7ft	Beautiful upright magnolia for a small garden. Deep pink-purple goblet flowers emerge in late spring from furry buds on bare stems.	Magnolias are woodland shrubs and trees which thrive on clay soil as long as it is not too dry in summer. Most prefer acid/neutral soil.
Mahonia nitens 'Cabaret'*PBR* AGM 1.5 × 1m/5ft × 39in	Holly-like dark, shiny leaves. Trusses of unusual orange-red buds open to yellow-orange flowers in late summer to autumn. Blue-grey berries.	
Pyrus salicifolia 'Pendula' 8 × 8m/26 × 26ft	A small tree for a confined space. It has silver leaves and a weeping habit that looks wonderful near water. Fruits inedible.	Prune from underneath, taking out whole branches to thin, rather than giving it a haircut all over, which will spoil its natural shape.

 Sun   Shade   Sun/partial shade ⠿ Any

PLANT	DESCRIPTION	NOTES
Solidago rugosa 'Fireworks' AGM 1 × 0.7m/39 × 27in ⊙	Sprays of bright yellow wands July–September. 'Naturalistic' in style and loved by bees.	Many people don't like solidago, but this one is more delicate and well-behaved than most.
Sorbus ulleungensis 'Olympic Flame' AGM 4–8 × 2.5–3m/ 13–26 × 8–10ft ⊙	A small, upright tree with cream flowers late spring, amazing autumn leaf colour. Red berries, loved by birds, persist into winter.	A great tree for a small garden with three seasons of interest, it really earns its place.
Viburnum × bodnantense 'Dawn' AGM 2.5 × 1.5m/8 × 5ft ◑	Semi-evergreen rounded shrub with clusters of pale/mid-pink highly scented flowers in Jan–February.	Look out for nibbles on the leaves from viburnum beetle. Requires little maintenance except occasional light pruning after flowering to keep in shape and size.

SAND/SILT

PLANT	DESCRIPTION	NOTES
Albizia julibrissin f. *rosea* AGM 4 × 3m/13 × 10ft ⊙	A bushy, broad-crowned tree requiring a sheltered spot.	I love the lightness of this tree, with its finely cut leaves and fluffy flowers – very unusual.
Artemisia 'Powis Castle' AGM 60 × 90cm/2 × 3ft ⊙	Fast-growing silver-leaved evergreen shrub.	An invaluable finely-cut, silvery-leaved shrub grown for its foliage rather than flowers. Has a tendency to get leggy, so, like lavenders, cut back hard in late spring, but never into old wood.

Buddleja 'Lochinch' AGM 2.5 × 2.5m/8 ×8ft ◐	Fast-growing deciduous shrub with silver leaves and grape-like panicles of scented mauve flowers with tiny orange eyes.	Buddlejas do not deserve their bad reputation. If pruned hard in March, they can be kept in their space and will produce long-lasting panicles with enviable wildlife value. I like to train them on to a few main stems to become tree-like.
Caryopteris × *clandonensis* 'Sterling Silver' AGM 1.2 × 1.2m/4 × 4ft ☉	A rounded shrub with glaucous, aromatic leaves and ghostly pale blue-purple flowers July–September.	Loved by butterflies, bees and moths. Can get a bit straggly, trim as required in spring.
Choisya × *dewitteana* 'White Dazzler' AGM 1 × 1m/39 × 39in ◐	Flowers April–May and often again in the autumn.	Really useful, compact evergreen with finely cut leaves and a profusion of white blossoms.
Cistus × *argenteus* 'Silver Pink' AGM 1 × 1m/39 × 39in ☉	Tissue-paper, pale pink, open flowers June–August.	A very useful evergreen mound with silvery foliage and a profusion of summer flowers. Cistus are fairly short-lived shrubs, but worth it and easy to replace. Can get straggly, so give a light trim after flowering.
Convolvulus cneorum AGM 0.6 × 1.9m/2 × 11⅓ft ☉	Silver-leaved evergreen with lax habit that will trail a little over walls or pots. Trumpet-like pure white flowers May–July.	Does well in a generous pot mulched with shingle.

☉ Sun　● Shade　◐ Sun/partial shade　⋮⋮ Any

Eschscholzia californica 40 × 40cm/16 × 16in ⊙	Trouble-free annual that self-seeds merrily if happy with your free-draining, sunny site. Beloved of pollinating insects.	Invaluable for pops of bright colour here and there. There are now many coloured cultivars to choose from –'Rose Chiffon' has the AGM.
Juniperus communis Upright, conical 8 × 4m/26 × 13ft In exposed conditions it can develop a stunted habit. ⊙	An evergreen shrub or small tree. Needle-like leaves and edible, aromatic berries used for making gin.	Rather than choosing an AGM juniper, I would like to highlight the UK native. This species is now endangered and there is a campaign to increase it.
Salvia (Perovskia) 'Blue Spire' AGM 1.2 × 1m/4ft × 39in ⊙	A robust sub-shrub with hazy, pale-purple mass of flowers amid glaucous foliage. Flowers August–September.	I also use 'Little Spire' for a more compact version.
Santolina rosmarinifolia subsp. *rosmarinifolia* 'Primrose Gem' AGM 40 × 60cm/16 × 24in ⊙	Bright green feathery mounds of evergreen foliage. With pale, cream-yellow pompom flowers July–August.	This slow-growing shrub is quite difficult to find, but worth searching out. It can be clipped in late spring to maintain a neat evergreen mound. Much subtler than the usual silver-leaved cotton lavender.
Verbena bonariensis AGM 1.5 × 0.3m/5ft × 11in ⊙	Long wiry stems support pompoms of deep purple flowers that sway in the breeze June–September. Deservedly popular.	Happily self-seeds around. If it likes your soil, you will find yourself donating seedlings to many friends.

ACID-LOVING

PLANT	DESCRIPTION	NOTES
Acer palmatum 'Sango-kaku' AGM 4 × 2m/13 × 7ft	It starts with bright salmon young stems in spring, with new, pink foliage. As the stems darken to reddish tones, the leaves become greener, then yellow in autumn.	So many acers to choose from, but this Japanese acer is my all-time favourite for its extended seasons of interest.
Calluna vulgaris 'Dark Beauty' AGM 20 × 40cm/8 × 16in	Evergreen. Cerise pink flowers darken to rich ruby red August–October. Moisture-retentive soil.	Deer resistant. Does not like to dry out. Tidy in spring by cutting back to within a couple of centimetres of the old growth.
Camellia sasanqua 'Crimson King' AGM 40 × 60cm/16 × 24in	Scented, deep pink-red open flowers October–November. Shelter from cold winds and also early morning sun when frosty.	Sasanqua camellias flower in autumn/winter. The small, simple flowers often have a pleasant earthy, mossy scent. Like all camellias the blooms may brown if touched by sunlight after morning frost. I like to train them against a wall, where they will be protected and also the flowers can be better appreciated.
Glandora prostrata 'Heavenly Blue' (previously known as *Lithodora*) 30 × 60cm/12 × 24in	Prostrate evergreen shrub with intense pure blue flowers May–August.	Young plants need extra care: shelter from cold winds, drying out and waterlogging until established.

 Sun   Shade   Sun/partial shade   Any

Hamamelis mollis 'Wisley Supreme' 3 × 3m/10 × 10ft ☉	Yellow autumn foliage. Sweetly scented pale yellow, spider-like flowers on bare stems December–February.	Dislikes drying out, especially when young. They are often grafted on to vigorous rootstock, so do not bury the union (the lumpy part of the trunk where the two plants have been joined together) when planting and take off any suckers growing from beneath the soil.
Hydrangea paniculata 'Pinky Winky' AGM 1.5 × 1.5m/5 × 5ft ☉	Buds open white and turn pink August–September.	Hydrangeas have had a resurgence in popularity, especially with the flamboyant paniculatas. Hard pruning in spring develops larger clusters of flowers on shorter stems. All hydrangeas dislike drying out.
Kalmia latifolia 'Freckles' AGM 1.8 × 1.8m/5.9 × 5.9ft ☉	Candy pink flowers with 'freckles' of wine-red. May–June Moist soil. Hates to dry out in summer.	Rarely seen, these acid-loving shrubs have distinctive clusters of flowers like tiny sugar-icing cups. They are rather fussy and hate too much nitrogen, waterlogging, drying out, cold winds or being moved. Good luck!
Liriope muscari AGM 40 × 40cm/ 16 × 16in ☉	Evergreen, leathery, grassy foliage. Spires of purple flowers August–November, followed by black berries.	A very useful, shade tolerant, robust and unfussy evergreen. Its tendency to look tatty can be improved by cutting off brown leaves to ground level in spring.

Magnolia 'Gold Star' AGM ◉	Pale yellow buds opening to cream/white, open, star-shaped flowers March–April. Sheltered spot.	Relatively fast growing to compact size, suitable for smaller gardens. Flowers are more frost resistant than many magnolias.
Parrotia persica 'Vanessa' 7 × 4m/25 × 13ft ◉	Spidery, red, witch-hazel-like flowers in winter. Moist soil. Compact and upright habit suitable for a smaller garden.	Known for its fabulously vibrant leaf colours. Young leaves are bronze/red, slowly changing to red, yellow and orange, and the tree often has many shades at the same time.
Pieris 'Forest Flame' AGM 2.5 × 1.5m/8 × 5ft (or even larger with time) ◉	Clusters of white, lily-of-the-valley-like flowers, attractive to bees March–May.	Amazing leaf colours for an evergreen. Bright pink/red young leaves change slowly to green. A long-lived and reliable shrub.
Rhododendron viscosum AGM 1.5 × 1.5m/5 × 5ft ◉	Scented white trumpets June–July.	This is unusual for a deciduous azalea as it prefers damp soil. It is highly fragrant, well worth a spot in the garden, despite its relatively short flowering time.
Vaccinium corymbosum 'Dwarf Blue Sapphire' 60 × 60cm/24 × 24in ◉	Pollinator attractant flowers March–April, followed by fruit in summer.	One advantage of having an acidic soil is that you can grow your own blueberries and cranberries. Most blueberry bushes take up a lot of space, but here is a recent dwarf introduction which you could also grow in a pot or raised bed.

 Sun   Shade   Sun/partial shade   Any

CHALK TOLERANT

PLANT	DESCRIPTION	NOTES
Acer campestre 'Evenley Red' 8 × 6m/26 × 20ft ☉	Small, rounded tree suitable for small gardens. This cultivar has bright red autumn colour.	*Acer campestre* is a UK native, with warm buttery yellow leaves in autumn. It does well on chalky soils.
Erica × darleyensis 'Phoebe' 40 × 60cm/16 × 24in ☉	Long-flowering chalk-tolerant heather with strands of tiny pale pink bells from November–April. Shelter from cold winds and also early morning sun when frosty.	After flowering, cut down to within a couple of centimetres of old growth to tidy. Remember when planting to leave enough room for substantial mounds of evergreen foliage to develop.
Hydrangea 'Preziosa' AGM 1.5 × 1.5m/5 × 5ft ◐	Sterile pink flowers darkening to a wine colour August–September.	Young bronze leaves darken to red/black. Tolerates chalky soil as long as it has plenty of organic matter and does not dry out.
Magnolia × loebneri 'Leonard Messel' AGM 6 × 6m/20 × 20ft ☉	Rounded shrub with pale pink, open flowers in April.	Prefers acid soil, but will tolerate chalk, as long as it has enough moisture and depth of topsoil.
Rhododendron 'Taurus' Inkharho Group AGM 1.5 × 1.5m/5 × 5ft ◐	Deep red, showy, typical rhododendron flowers on big-leaved, evergreen shrub April–May. Moisture-retentive soil: pH 4.5 to 7.5.	This new group of lime-tolerant rhododendrons is grafted on to a rootstock from a plant found growing happily in a lime quarry. Feed regularly and keep moist.

Plant	Description	Notes
Salvia × jamensis 'Hot Lips' AGM 1 × 1m / 39 ×39in ◑	Masses of white and red flowers, which are attractive to pollinators July–October.	A curious, trouble-free and popular shrubby sage. Keep tidy by cutting back by one-third in spring.

CHALK LOVING

PLANT	DESCRIPTION	NOTES
Achillea 'Desert Eve Terracotta' 0.5 × 0.5m/20 × 20in ☉	A long-flowering perennial (May–July), with flat-topped clusters of rich terracotta flowers, fading to creamy yellow.	Species achilleas (or yarrows) are native to UL chalk grasslands, so cultivars do well on chalk or any well-drained soil. Ideal for cottage gardens and also prairie-style plantings with ornamental grasses. Deer and rabbit resistant.
Agapanthus 'Northern Star'[PBR] AGM 1 × 0.5m/39 × 20in	A hardy cultivar, 'Northern Star' has striped star-shaped flowers in July. It enjoys a well-drained soil of any type and is drought tolerant once established.	A fast-growing climber which is great for covering unsightly walls, even in shade. It has a subtle, but lovely chocolate smell, hence the name 'chocolate vine'.
Akebia quinata 'Alba' 10 × 2m/33 × 7ft ⋮⋮	A semi-evergreen twining climber with fresh green leaves that are bronze in spring and flush purple in autumn. In March–May it has clusters of tiny cream flowers with purple-red centres.	A fast-growing climber which is great for covering unsightly walls, even in shade. It has a subtle, but lovely chocolate smell, hence the name 'chocolate vine'.

☉ Sun   ● Shade   ◑ Sun/partial shade   ⋮⋮ Any

Campanula portenschlagiana AGM 20 × 40cm/8 × 16in ◑	Small blue-purple bells on creeping mounds of evergreen foliage June–August and intermittently.	Campanulas originally hail from chalk grasslands. This creeper will fill cracks and edge planters, but is not invasive.
Clematis 'Fujimusume' AGM 1.5 × 1m/5ft × 39in ◑	Very pale purple, large open flowers 20cm/8in diameter. Shelter from cold winds and also early morning sun when frosty.	Pruning for Group 2 clematis can get complicated, but an easier way is to cut the whole plant down to 30cm/12in every few years, and accept it will have reduced flowering that year.
Dianthus carthusianorum 60 × 20cm/24 × 8in ☉	Evergreen grassy leaves with wiry flower stems bearing small clusters of magenta pink flowers June–September.	Dianthus love chalk. The Carthusian dianthus is deservedly popular. It has not been awarded an AGM, but I am making an exception to include it, because it is invaluable in creating a cottage or informal naturalistic style in your planting.
Euonymus europaeus 'Red Cascade' 3 × 2.5m/10 × 8ft AGM ◑	A shrub or small tree with insignificant flowers following by curious berries with bright pink petals/bracts and orange centres. Amazing autumn colour of fiery, translucent red leaves.	Cultivar of a UK hedgerow native, with poisonous berries (European spindle). Birds love the berries. Autumn colour is brightest if planted in sun.
Fuchsia 'Whiteknights Pearl' AGM 1.5 × 1.5m/5 × 5ft ◑	A deciduous shrub that will tolerate shade and is very useful for adding height and colour to the back of a border. 'Whiteknights Pearl' has slender pale pink droplet flowers and mid-green deciduous leaves.	Fuchsia berries are edible (check your own tolerance before eating).

Iris germanica 'Jane Phillips' (tall bearded iris) AGM 90 × 30cm/36 × 12in ☉	Pale purple, lightly scented typical iris flowers on tall stems, with sword-like leaves. One of the earlier flowering bearded irises May–June.	Bearded irises love sun and chalk. To promote flowering, keep the fleshy rhizomes exposed above the soil surface where they can bake in the sun. Feed after flowering and reduce leaves by one-third in autumn to prevent root-rock.
Lavandula 'Grosso' 0.7 × 1m/27 × 39in without flowers ☉	6cm/2½in long, dark purple flowers above evergreen glaucous foliage July–September.	One of the latest and largest varieties, 'Grosso' is grown for harvesting of fragrant lavender oil. It does not have an AGM, but is still the most grown lavender in the world.
Penstemon heterophyllus 'Heavenly Blue' 0.5 × 0.5m/20 × 20in ◑	Semi-evergreen perennial, flowering May-October. Loved by bees.	One of my favourite long-lasting perennials, that has a shocking electric blue flower, with tones of purple of pink. Can be propagated by dividing in spring or taking cuttings.
Vitis 'Fragola' (Strawberry grape) 8m/26ft tall untrained ☉	Small, pink dessert grapes on a vine with beautiful yellow autumn foliage.	For better fruiting, divert energy to the grapes that your vine would otherwise put into leaves by pruning.
Weigela 'Florida Variegata' AGM 2.5 × 1.5m/8 × 5ft ◑	A deciduous arching shrub with variegated leaves. Pink trumpet flowers in May/June.	A classic 'English cottage garden' shrub, which is invaluable for plugging the early spring gap in flower colour.

☉ Sun   ● Shade   ◑ Sun/partial shade   ⠃ Any

SOIL TROUBLE-SHOOTING

HOW DO I MANAGE FLOODING?

KENT RAIN GARDEN

———

Above: Parts of the garden and surrounding farmland can lie under water for several months of the cold winter.

Signs of traditional farmland flood-management techniques can be a clue for gardeners that the area is likely to flood. I always keep an eye open for boundary ditches when I survey a new site, because they are often dug by farmers to drain their land. In this Kent garden, the drainage ditches used to flow out across the surrounding fields, but had become silted up over time. Blocked ditches, heavy clay soil, being at the bottom of the slope and an ornamental pond with nowhere to drain to all contributed to seasonal flooding of this garden.

My client suggested a rainwater garden, and being an enthusiastic gardener and wildlife supporter, he was excited about the possibilities of a biodiverse and environmentally resilient planting with ornamental interest.

A large field of grass drains down to the ditch and the pond overflows into it in winter, so much so that the area becomes impassable for weeks at that time. My starting point was to make it accessible with a decking boardwalk and then to plant trees and shrubs which would not only be happy in the wet, but also hold back some of the downpours, with their wide-reaching, thirsty root systems.

Once these main structural, woody plants were in place, I added an evergreen grass to create a winter covering and constant backdrop for the ornamental planting. Then, finally, the ornamental and wildlife-friendly flowering plants for summer interest. These had to be tolerant of standing water in the dormant winter months and also drier ground conditions in summer (see Plants for Rain Gardens, pages 116–17).

HOW DO I DEAL WITH EROSION CAUSED BY FLOODING?

RIVERSIDE PLOT

<u>Top</u>: The previous owner shored up the eroded riverbank with concrete and old scaffolding poles.

<u>Above</u>: Our new garden design incorporated plants adapted to survive many weeks underwater.

Several years ago, we were asked to renovate a derelict riverside plot to create a naturalistic space that did not look like a garden, but segued seamlessly into the surrounding Area of Outstanding Natural Beauty. The site had been a neglected and hazardous dumping ground for years, but the real challenge was that it floods every spring and it can stay underwater for several weeks. The banks were eroded and undercut, and receding floodwaters had left a thick layer of debris and detritus.

Upon analysing the site and its soil, I discovered four main effects of the flood water:

- The detrimental effect of high water levels on the plants themselves and the need for plants adapted to the situation.
- The damaging fast flow of river water across the site.
- The eroded banks to the island site.
- The silt and rubbish left after the water subsides.
- Old and dead trees which needed removing, but were probably contributing to holding the soil together.

Having cleared the site of rubbish, the next step was to manage the trees. Most of them were 'crack willows' which have a tendency to crack apart and fall over. They love watery habitats and being fast-growing they will quickly knit soil together with their roots. I did not want to remove them, but several were falling over into the water. We kept what we could and immediately started planting new fast-growing, moisture-loving plants to replace them.

The banks of the island had been undercut by fast-moving currents and over the years various 'Heath Robinson' attempts to shore it up with scaffolding poles and chunks of concrete had failed. Having consulted with the water authorities, I designed a new profile for the water's edge, to be constructed with biodegradable, planted marginal 'mattresses'. By the time the hessian material of the mattresses rot, the roots of the plants have knitted the soil of the new riverbank together.

I decided to use mainly deciduous plants because evergreens would be a trap for floating detritus, especially in the early spring floods. I also had a large proportion of cornus, salix and elder shrubs that all benefit from regular spring pollarding, which could become part of the annual clear-up.

There are strict regulations about building any structures on a flood-plain that might interfere with the flow of water. As a result, this design had no built structures; all the height, structure and interest was achieved with planting. Although the brief was for a naturalistic and wildlife-friendly garden, I did not think it necessary to restrict to UK natives only. The plants were carefully chosen to withstand floods, support wildlife, require minimal maintenance and have a native 'look' and ornamental value.

Top: The cleared site and re-configured riverbank.

Above: A year after planting, the native and naturalistic plants have withstood their first spring flood.

HOW DO I CONTROL PESTS AND DISEASES?

SUSSEX LAVENDER FIELD

Top: We added grit to aerate the heavy clay and planted the lavender on mounds to aid drainage.

Above: The lavender field established well, but some plants succumbed to *Phytophthora* infection.

Beth Chatto is attributed with first coining the phrase 'right plant, right place', and it is a mantra that has served me and other garden designers well for the past few decades. However, every now and then one is tempted to set aside what one knows to be true and this garden was one such case for me. My client, having holidayed in the South of France, had what I call a 'post-holiday, ouzo-moment', and reluctant to return to the grey mizzle of England, she hatched a plan to plant her own Provencal lavender field. The problem was that she had not the free-draining warm soils of the Mediterranean, but the cold, claggy clay of the Sussex Weald.

Invoking my traditional knowledge of how to 'improve' a clay soil, we added tonnes of grit to the would-be lavender field and mounded the soil into ridges to aid drainage.

Putting aside the environmental costs of importing soil improvers and cultivating the land, the project was on the whole a success. However, after a year or two, in discrete spots we started to lose plants to a fungal-looking wilting disease. Diagnosis from a plant laboratory soon confirmed *Phytophthora* root rot. This common condition is caused by soil-borne fungus-like organisms that can attack woody plants in particular if they are not well drained enough.

Advice to combat *Phytophthora* is to remove contaminated soil if the condition is recent and localised, improve soil drainage, plant resistant varieties (not lavender) or dig a pond!

Plants resistant to *Phytophthora* include:

1. *Acacia pravissima*
2. *Carpenteria californica*
3. *Catalpa bignonioides*
4. *Deutzia* types
5. *Gleditsia 'Skyline'*
6. *Koelreuteria paniculata*
7. *Nandina domestica*
8. *Pyrus communis*
9. *Trachelospermum jasminoides*
10. *Weigela* types

Apart from *Phytophthora*, the most common soil-borne pathogens in the UK are mostly troublesome for vegetable growers. This includes onion white rot and brassica club root.

Plasmodiophora brassicae, club root microorganisms, can remain in the soil for decades, so I am afraid if you have it, it is difficult to get rid of. Increasing the pH of the soil by adding lime may help, so this is one instance (beyond the scope of this book) where trying to alter the chemical composition of your soil may be advisable. Plasmodiophora causes massive swelling of roots and severely retarded growth, usually in the warm summer months. The best strategy is to try to avoid your soil becoming affected in the first place by keeping it well drained and using only reputable plant suppliers or growing from seed. There are also some resistant cultivars available.

Onion white rot, *Stromatinia cepivora*, is another long-lived soil organism, this time a fungus. Like club root, once you have it, it is difficult to eliminate, so prevention, particularly with good soil

hygiene, is essential. Avoid bringing other people's soil into your garden, particularly that of near neighbours, like fellow allotment-holders. This disease strikes in cooler wet months, causing wilting, yellow foliage and fluffy growths around the base of the onion.

Top: Tomato blight on tomato fruits, caused by *Phytophthora infestans*.

Bottom: Brassica club root in sprouts.

HOW DO I PREVENT SOIL COMPACTION WHEN BUILDING?

SOIL PROTECTION ON A CONSTRUCTION SITE

―――――

When a landscaping, farming or building project is underway, there is often a deadline. I have discussed why there are good and bad climatic conditions and times of year to work soil, and that there is never a good time to compact it. However, that would be in an ideal world. In reality, weather is unpredictable and time is money.

For these reasons, even if the risks to soil structure are understood, often work will proceed in wet and cold conditions, and corners will be cut.

In Britain there are national building guidelines to protect soil which should be followed. Architects, structural engineers, landscape designers and project managers should be aware of these and the necessary operating parameters included in specifications and contracts.

Overlooking or circumventing these guidelines is often done in the name of saving time, but in the long run a compacted soil can result in later unanticipated costs. Soil damage and compaction can cause flooding, erosion and failure of newly planted trees and shrubs.

Time efficiency is not the only reason for failing to avoid compaction of soil. I think that people in the building industry and people in landscaping have a fundamentally different perspective on soil that needs to be addressed. For people focused on building safe structures for us to inhabit, 'dirt' is the starting point for a secure foundation. Compacting that sub-grade, often by mixing it with cement, is a tried and tested method for creating a stable base from which to start building up. Soil is seen as a freely available construction material and the subtle distinction between topsoil and subsoil is often overlooked.

The perspective of the garden-maker or environmentalist is the opposite. For us, soil is a delicate and vulnerable complex ecosystem. It is far better to prevent damage than to try and fix it retrospectively.

On a construction site, soil (and the roots of existing trees) should be protected by involving the whole team in agreeing the borders of the construction site area and then fencing that area with semi-permanent fencing, such as 'Heras' steel mesh. Roads and pathways should be kept to a minimum and routed away from trees. The soil beneath routes should be protected with cellular mesh to distribute the weight.

Soil protection on a construction site requires a whole team approach. If that site belongs to you, then, as the client, your voice can go a long way towards focusing minds on the preservation of your land.

IS IT OKAY TO IMPORT TOPSOIL?

IMPORTING TOPSOIL TO A SOIL DEPRIVED SITE

Plants have adapted to survive all over the globe, often thriving in extreme conditions and with seemingly hostile soils. Knowing what we now know, we should not reach for a bottle of fertiliser to feed our plants, but instead look to natural processes to 'feed' the soil – adding organic matter to enable the soil to better hold on to any available nutrients and deal with difficult environmental conditions.

Some of the sites that I work on have suffered severe 'soil violence', often to the extent that there is not enough native soil left alive. Given time, organic matter can actually reinvigorate a seemingly dead soil; it likely contains the microorganisms, eggs and larvae to do just that.

However, sometimes we need to import topsoil to build up the necessary levels to create a garden. You might think that topsoil would have everything that a plant needs for healthy growth, but that very much depends on its source and specification. Topsoil or 'earth' for a gardener is a complex, balanced, living ecosystem with biological as well as chemical and physical parameters that support biodiversity. Topsoil or 'dirt', for a builder or engineer, can be an inert material for supporting artifcial structures.

Previously it was difficult for a consumer to know exactly what they were getting when buying imported topsoil, but recent improvements in the British Standards for specifying soils have gone some way to helping achieve consistency.

Imported soil can be natural soil which is dug up from one site and taken to another, or it can be manufactured by blending different constituents. The BS3882:2015 standard soil is multipurpose and has set parameters for its properties and nutrients, so you can be sure of a minimum good standard.

Your supplier should be happy to give you information about the source of the soil and also a certification of its chemical properties. A BS3882:2015 topsoil should have minimum nutrient levels for general landscape use and plant growth. However, it is best to see imported topsoil as a starting point; you are likely to need to begin a programme of soil improvement by adding organic matter over time in order to achieve reliable soil performance to support the plants in your garden. Importing is always disruptive – work with what you have on site wherever possible.

HOW DO I ESTABLISH A GARDEN ON A NEWBUILD SITE WITH DEAD SOIL?

RAISED BEDS ON A NEW BUILD SITE

If you are starting a garden on a newbuild site, you may well have the unholy trinity of compacted native soil, poor drainage and a skim of poor-quality imported topsoil spread over the ground to disguise the layer of builder's rubble beneath.

If the situation is bad and complex, then I suggest that you seek professional help. However, in most cases, with careful planning and investment of time – and I am afraid some money – you will be able to improve the situation to a point where you can establish your new garden.

With this example, we had poorly draining compacted clay subsoil with very little topsoil. The garden pooled with water in winter and the lawn was on its last legs. It seemed impossible for my clients to grow any plants, except a couple of die-hard bay trees.

The first thing I noticed was that the whole garden was enclosed by brick walls, so when rainwater ran off the surface of the

compacted clay, it had nowhere to drain to. It couldn't travel down through the soil layers and it couldn't run off elsewhere. There were no plants to speak of to 'drink' the water, and the lack of plants meant there was no chance of building up the organic matter content of the soil any time soon.

Digging an exploratory pit, I found a thin layer of topsoil over a deep subsoil layer of impenetrable clay, dotted with old bricks and clumps of concrete. I wasn't too worried about the odd half-brick – plants don't mind that – but my first step was to dig land drains to a soakaway to help water drain away. However, on heavy clay, there is only so much land drains can do because it takes an age, if at all, for the water to travel through the soil and away from the 'soakaway'.

So, belt-and-braces were required; I couldn't rely on land drains to solve the flooding issue – we would need to rise above it. I designed a garden of raised beds instead.

GLOSSARY

AGM The RHS' 'Award of Garden Merit', given to plants that have performed well in garden trials and so are deemed 'garden worthy' and reliable.

alkaline Having a pH greater than 7 and properties opposite to an acid.

anaerobic Without oxygen.

aquifer Permeable underground layers of rock that store groundwater and allow it to flow.

attenuation To reduce the effects of flood water by storing it temporarily, to then be released gradually and in more controlled manner.

bedrock The lowest solid layer of rock, from which soil layers are derived.

biodiverse Having a wide range of plants, animals and other living organisms.

blood-brain barrier A permeable barrier between the blood and the fluid of the brain, that filters particles.

carbon A non-metallic element, which can be found in inorganic materials such as diamonds, graphite, coal and petrol, as well as in living and organic organisms.

carbon sequestration Isolating and storing carbon.

carbon sink Something that stores carbon by absorbing more than it releases.

chlorophyll The green pigment found in green plants and cyanobacteria, which absorbs light to enable photosynthesis.

chlorosis Loss of green colour in leaves as a result of reduced chlorophyll.

comfrey A plant from the borage family, that is high in nitrogen and potassium.

DEFRA UK government's Department for Environment, Food, and Rural Affairs.

denitrification The process of converting nitrogen compounds, like nitrates, into nitrogen gas and nitrogen oxides, and releasing them into the atmosphere.

DNA Deoxyribonucleic acid is the molecule that carries the genetic information of an organism.

double-digging A traditional digging method of turning over the soil to a depth of approximately 60cm (2ft), or two spades' depth.

ecosystem A complex network of different organisms, interacting with their local physical environment and its constituents

ericaceous Plants that grow best in soil of pH 6 or below or the acidic soil that supports them.

exudate A substance secreted.

fungal hyphae Tubular cell-like parts of fungi that allow them to interact closely with soil particles and organisms.

geotextile Permeable natural or synthetic material used in civil engineering, building, gardening or landscaping.

green manure Crops grown specifically to improve the fertility or structure of the soil.

grit Small pieces of rock or sand. Horticultural grit is usually angular, of approximately 3mm in diameter and used to improve drainage of soil or compost

gypsum A mineral (hydrous calcium sulphate) used to fertilise the soil.

hoeing To disturb the surface of the soil with a hoe, to disrupt the growth of weed seedlings and aerate the soil surface.

hoggin A mixture of stones of different sizes, with enough clay particles to act as a natrual binder when compacted and wet.

hydroponics A growing technique using a water-based nutrient solution, rather than soil.

ion/nutrient ion An atom or group of atoms that has a tiny electrical charge and so can be attracted or replled by other charged particles.

leachate A solution that forms when a liquid travels through a solid and washes out elements from that solid material.

lime A white caustic alkaline substance containing of calcium oxide; however, gardeners often use the term 'lime' to mean chalk.

marginal plants Plants adapted to grow in the shallow areas at the edge of a body of fresh water.

microbiome A community of microorganisms that grow in a particular environment.

microorganism An organism that can only be seem through a microscope.

mineral A solid, naturally occurring inorganic substance.

minibeast A colloquial term for small invertebrates.

mucilage A viscous secretion or gelatinous fluid.

mulch A loose organic or inorganic covering on the soil surface.

mycorrhizae Ubiquitous fungi that grow in the soil, in close, often symbiotic, association with plant roots.

no-dig A gardening method which avoids digging, breaking up or turning the soil.

organic matter Carbon-containing matter derived from living organisms and often referring to the remnants of rotting, or decomposed material.

ozone layer A protective layer in the stratosphere, which encirces the earth and aborbs harmful ultraviolet radiation.

perennial A plant which regrows each growing season. Its leaves die down in the winter months, but its roots remain dormant and viable underground

permeable A material which allows liquids or gasses to pas through it.

pH 'Potential of hydrogen': a measure of acidity or alkalinity of a solution or gas.

photosynthesis The process by which plants use sunlight, water and carbon dioxide to create oxygen and energy in the form of sugar.

Phytophthora A large group of pathogenic organisms that resemble fungi.

pioneer species The first organisms to colonise a new or freshly cleared area.

pore A small hole or gap.

profile pit A vertical-sided hole, dug to reveal the profile of soil horizons.

rhizosphere The narrow area of soil surrounding and influenced by a plant root.

soil horizon A distinct layer within a soil, which has different physical and chemical properties to the layers above and beneath.

soil structure The arrangement of soil particles into different sized aggregates or 'peds'.

soil texture The proportions of soil particles of different sizes: clay, silt and sand.

solution A liquid mixture with another component (solid or gas) evenly distributed within it.

subsoil The soil layer beneath the topsoil, which has similar mineral particles, but less organic matter.

SUDS (Sustainable Urban Drainage System). A drainage system that mimics natural water flow through the landscape and reduces the flow of rainwater into pipes and sewer systems, to be discharged into water systems.

symbiosis A close association of two organisms to mutual benefit.

topsoil The top layer of soil which has mineral particles and a relatively high proportion of organic matter.

turves Lawn grass and the shallow layer of topsoil holding its roots.

wormery A container housing worms being bred or used for composting.

INDEX

A

acer 48

acid soil 20, 28, 40–2, 44, 48, 62, 75, 77, 78, 131
　　suggested plants 135–7

aeration 15, 17, 27, 32, 34, 43, 46, 56, 65, 90,
　　96, 100, 120, 147, 153

aerobic 26, 27, 77, 78

aggregation 14, 15, 17, 25–8, 48, 56–9, 96, 98,
　　120, 154

AGM (Award of Garden Merit, RHS) 126–8,
　　130–41, 152

air plants 66

alkaline 20, 42, 44, 76, 152, 153

anaerobic 26, 27, 30, 45, 50, 78, 152

annuals 120, 121, 134,

aquaponics 67

aquifer 26, 96, 152

ash 36

attenuation 110, 111, 152

B

bacteria 13–15, 18, 19, 22, 25–7, 30–2, 76, 78,
　　79, 82, 121, 122, 152

barley 122

bedrock 17, 54, 55, 152

beech 82

biodiversity 12–23, 28, 29, 34, 68, 112, 113,
　　126, 144, 150, 152

biogeochemical cycles 7, 13, 15, 22–24, 30–7,
　　62, 93, 96, 110

birch 48, 124

blood-brain barrier 34, 152

Bokashi bin fermentation 77–9

box 36

bracken 48

brassicas 148

C

cacti 65

calcium carbonate, *see* chalky soil

camellia 48

carbon 7, 13–15, 18, 22, 24, 28–30, 32–4, 37,
　　77, 106, 152, 154
　　cycle 32, 33, 36
　　sequestration 29, 32, 152
　　sink 29, 37, 152

chalky soil 20, 40, 41, 44, 48, 50, 68, 84,
　　85, 153
　　suggested plants 138–41

Chatto, Beth 45, 46, 68, 86, 147

chemical characteristics of soil 60, 61

chicory 28

chlorophyll 20, 121, 152

chlorosis 21, 44, 152

clay 12, 14, 16, 17, 20, 26, 27, 40–7, 50, 52, 60,
　　62, 84, 98, 99, 120, 121, 126, 144, 147, 151,
　　153, 154
　　suggested plants 120, 121, 130–2

climate change 13, 15, 24–38, 86, 126

clover 28

club root 148

comfrey 20, 21, 88, 120, 126, 152

common groundsel 48

compaction of soil 15, 26–8, 30, 45, 46, 48,
　　56, 62, 96, 100, 101, 109, 120, 121, 126, 149,
　　151, 153

compost 20, 21, 28, 29, 65, 66, 72–83, 88, 90, 92, 102, 104, 106, 109, 126, 153, 154
container gardening 74, 100, 101, 104, 113, 114
cornus 146
cotton clothing 49
creeping buttercup 27, 48
creeping thistle 48
cyanobacteria 13, 18, 32, 152

D
dandelion 120
decompaction 120, 126
decontamination of soil 122
deforestation 25
DEFRA (Department for Environment, Food and Rural Affairs) 37, 152
denitrification 30, 152
devil's bit scabious 48
disease control 13, 36, 40, 66, 86, 92, 147, 148
    suggested plants 148
DNA (Deoxyribonucleic acid) 30, 152
double-digging 84, 152
drainage, of soil 16, 27, 29, 41, 43–5, 56, 65, 68, 76, 85, 95–106, 111–14, 134, 139, 144, 147, 148, 151, 153, 154

E
ecosystem 22, 29, 30, 34, 68, 111, 120, 149, 150, 153
elder 146
epiphytes 65, 66
ericaceous 20, 75, 153
    suggested plants 135–7
    see also acid soil
erosion 15, 22, 34, 37, 62, 92, 121, 124, 145, 149
exudate 25, 90, 153

F
farming 15, 25, 33, 34, 68, 74, 92, 144, 149
fertilising 20–3, 26, 30, 34, 37, 42, 68, 78, 79, 85, 88–92, 106, 150, 153

flooding 13, 15, 26, 27, 30, 41, 111, 112, 116, 126, 144–6, 149, 151, 152
fossil fuels 32, 33
fungal hyphae 14, 15, 100, 153
fungi 14, 15, 18, 19, 22, 23, 25, 27, 76, 79, 82, 93, 100, 130, 147, 148, 153, 154

G
geotextile 102, 114, 153
gorse 48
grasses for stabilising soil 124
gravel 85, 86, 92, 96, 102, 104, 113
green manure 28, 153
grit 16, 42–4, 47, 50, 65, 113, 147, 153
groundcover 25, 124, 126, 130
    suggested plants 124, 126
gypsum 42, 153

H
heather 48
hedges 15, 25, 130, 140
hemp 122
hoeing 84, 153
hoggin 96, 153
house plants 64, 65, 74
hügelkultur 106–9
hydroponics 66, 68, 153

I
impermeability, see permeability
importing soil 150
ions 45, 47

J
John Innes composts 74, 75

K
keyhole beds 104, 105
kitchen waste 72, 78, 85, 90, 101, 102, 104
knapweed 48

L

lavender 147
leachate 78, 79, 104, 153
leaf mulch 82, 83
legumes 18, 19, 121, 122
lime 41, 42, 44, 138, 148, 153
loam 40–3, 46, 47, 74, 84
    suggested plants 127–9
lucerne 28

M

macroorganisms 62, 79
manure 20, 28, 72, 74, 85, 90, 104, 109, 129
marginal plants 112, 116, 153
microbiome 18, 19, 153
microorganisms 7, 12, 13, 17, 18, 22, 30, 32, 45,
    62, 62, 79, 100, 122, 148, 150, 153
microplastics 34, 92
minerals 12–15, 17, 18, 54, 55, 62, 66, 85,
    153, 154
    mineral soil 14, 17
minibeasts 12, 13, 15, 23, 48, 64, 153
moss 48, 65, 66
mucilage 14, 18, 153
mulch 15, 20, 62, 72–83, 86, 92, 94, 106, 129,
    133, 153
mycorrhizae 18, 153

N

new builds 151
nitrogen 18–20, 22, 24, 27, 36, 44, 62, 68, 72,
    76, 77, 88, 106, 126, 136, 152
    cycle 30, 31, 36
    fixation 121, 122
    pollution 30
no-dig gardening 100–103, 154
nutrients 7, 13, 16–22, 28, 34, 43–8, 56, 62, 65,
    66, 68, 74–7, 85, 92, 104, 106, 124, 150, 153
    see also ions, minerals, nitrogen,
        phosphorus, potassium

O

oak 36
onion white rot 148
orchid 48, 65
organic matter 7, 12–17, 20–22, 25–28, 30,
    32, 33, 36, 42, 48, 50, 54–6, 60, 62, 66, 72,
    76, 77, 79, 82, 85, 86, 92, 101, 104, 106, 109,
    120, 121, 129, 138, 150, 151, 154
organic soil 14
ozone layer 30, 154

P

paving 95–8, 111
peat 14, 28, 29, 41, 65, 66, 74–6
perennial 88, 102, 109, 116, 117, 120, 121, 127,
    130, 139, 141, 154
permeability 15, 26, 41, 96, 98, 102, 112–14,
    152–4
pest control 13, 22, 36, 66, 86, 92, 147, 148
    suggested plants 148
pH 19, 20, 27, 34, 44, 60, 75, 138, 148, 152–4
phosphorus 18–20, 22, 24, 62, 79, 88, 122
photosynthesis 13, 18, 20, 22, 32, 33, 37, 44,
    121, 152, 154
Phytophthora 147, 148, 154
phytoremediation 122
pine 36, 48
pioneer species 124, 154
plantain 48
plant suggestions
    for acid soil 135–7
    for chalky soil 138–141
    for clay 120, 121, 130–2
    for loamy soil 127–9
    for pest and disease control 148
    for rain gardens 116, 117, 144
    for sandy soil and silt 132–4
    for saving soil 126
    for stabilising soil 124, 126
pollution 22, 30–4, 98, 122
poplar 122

pores 14, 15, 22, 26, 27, 56, 93, 111, 154
potassium 20, 44, 62, 88, 152
profile, of soil 17, 20, 54
    profile pit 17, 154

R

rain gardens 110–15, 144
    suggested plants 116, 117, 144
raised beds 77, 100–4, 111, 126, 137, 151
rhizosphere 18, 19, 122, 154
rhododendron 48
riparian, *see* marginal plants
rose 88
rubble 68, 151
rye grass 121

S

salad crops 66
salix 146
sandy soil 6, 14, 16, 20, 27, 40–3, 46–8, 50,
    52, 62, 68, 84, 85, 99, 111, 124, 154
    suggested plants 132–4
seasons of the year 62, 66, 93, 99
sedimentation 14, 52
shrubs 25, 75, 124, 127–141, 144, 146, 149
silt 14, 16, 27, 40, 43, 46, 52, 62, 144, 145, 154
    suggested plants 132–4
silver birch 124
slake test 58, 59
soil
    horizon 17, 54, 55, 84, 154
    saving, suggested plants 126
    structure 12–23, 25–33, 36, 37, 42–8, 56,
        58, 84, 85, 98–101, 120, 149, 154
    texture 16, 17, 27, 43, 44, 54, 56, 74, 75,
        82, 154
    types, *see* clay, sandy soil, silt
solution 45, 153, 154
sphagnum moss 66
stabilisation of soil 16, 25, 92, 124–6
    suggested plants 124, 126

stinging nettle 48, 88
strawberry 66
subsoil 17, 26, 27, 54, 55, 101, 149, 151, 154
SUDS (Sustainable Urban Drainage
    System) 98, 154
sugar beet 74, 122
sunflower 122
sweet chestnut 36
Swiss Cheese plant 65
switchgrass 122
symbiosis 13, 18, 22, 121, 153, 154

T

tea for compost 88
tomato 20, 74, 75, 88, 129, 148
topsoil 17, 25, 37, 50, 54, 55, 60, 74, 92, 101,
    102, 109, 124, 138, 149, 150, 151, 154
trees 13, 15, 25, 32, 36, 48, 75, 82, 92, 122, 124,
    126, 131–6, 137, 138, 140, 144, 145, 149, 151
trouble-shooting 144–51
turves 74, 109, 154

V

vegetables, growing 42, 48, 74, 75, 77, 84,
    88, 100, 101, 106, 148

W

water cycle 7, 12–18, 24–30, 36, 37, 41, 62, 64,
    66, 96, 110, 111, 122, 124, 154
watering 15, 65–7, 86, 90–8, 104, 106,
    109–15, 122, 129, 130
weeds 27, 48, 62, 74, 76, 78, 82, 84–6, 88, 92,
    98, 100, 102, 104, 109, 126, 153
white mustard 120–2
willow 122, 145
wormery 79, 90, 91, 154
worms 17, 22, 23, 26, 34, 48, 62, 72, 77, 79, 90

Y

yarrow 48, 121
Yorkshire fog grass 27

PICTURE CREDITS

ACKNOWLEDGEMENTS

To my clients who have given me the privilege of working on their land; each day learning more about the soil community that sustains it.

To A&P, who gave me the opportunity to delve deeper into the art and science of soil.

Quarto

First published in 2025 by Frances Lincoln,
an imprint of The Quarto Group.
One Triptych Place
London, SE1 9SH,
United Kingdom
T (0)20 7700 9000
www.Quarto.com

EEA Representation, WTS Tax d.o.o., Žanova ulica 3, 4000 Kranj, Slovenia.

Text © 2025 Juliet Sargeant
Photography © 2025 Matthew J. Thomas (www.matthewjthomas.com),
except where listed on page 159.

Published in association with the Royal Botanic Gardens, Kew.

A catalogue record for this book is available from the British Library.

ISBN 978-0-7112-8939-0
Ebook ISBN 978-0-7112-8940-6

10 9 8 7 6 5 4 3 2 1

Design by Michelle Kliem

Editor: Alice McKeever
Senior Editor: Michael Brunström
Editorial Assistant: Izzy Toner
Art Director: Isabel Eeles
Production Controller: Alex Merrett

Printed in China